DESIGN AND ANALYSIS OF EXPERIMENTS
in the Animal and Medical Sciences

VOLUME 3
APPENDICES

DESIGN AND ANALYSIS

in the Animal

THE IOWA STATE UNIVERSITY

OF EXPERIMENTS
and Medical Sciences

JOHN L. GILL

VOLUME 3
APPENDICES

PRESS / *Ames, Iowa, U.S.A.*

John L. Gill is professor of biometry in the Department of Dairy Science at Michigan State University. He received the M.S. and Ph.D. degrees in animal genetics and statistics at Iowa State University. Besides this book, he is author of many articles in the journals of his fields, and he has had long experience as a teacher of applied statistics and statistical consultant in the biological sciences.

© 1978 The Iowa State University Press
All rights reserved

Printed by
The Iowa State University Press
Ames, Iowa 50010

First edition, 1978

Library of Congress Cataloging in Publication Data

Gill, John L.
    Design and analysis of experiments in the animal and medical sciences.

    1. Zoology, Experimental—Statistical methods.
2. Medicine, Experimental—Statistical methods.
I. Title.

QH323.5.G55   591 .072   78-1045
ISBN 0-8138-0110-9

CONTENTS

PREFACE

Volume 3, appendices to the two volumes of *Design and Analysis of Experiments in the Animal and Medical Sciences*, provides tables and charts (App. A), glossary of symbols (App. B), and solutions to odd-numbered exercises (App. C). The need for frequent and easy use of these features with either volume dictated making them available separately.

The set of tables and charts grew out of the needs of biologists with whom I have consulted. The common tables of standard normal, Student's t, chi-square, variance-ratio (F), and Studentized range are more extensive than usual. Many others rarely found in applied texts are included. Percentage points are given for Hartley's F_{max} test for heterogeneous variances, Dunnett's t tests against a control, the Bonferroni t and χ^2 statistics for multiple comparisons of continuous and counted data, Hotelling's T^2 test (the multivariate analogue of Student's t), and Wilks's likelihood-ratio test (a multivariate analogue of F). Also given are orthogonal polynomial coefficients for three and four unequally spaced doses, the Shapiro-Wilk tables for testing normality, two tables of Grubbs and Beck for testing outliers, and 21 (11 completely new) power charts.

Instructions for use of the tables and charts (when not given here) can be found in appropriate sections of Vol. 1.

Tables of square roots and logs are not included because of the widespread availability of mathematics tables and pocket calculators with automatic functions.

The glossary is based on a set of notational conventions given at the beginning. These conventions were selected to conform as closely as possible with the recommendations of the Committee of Presidents of the Statistical Societies, while maintaining consistency throughout both volumes and staying within the bounds set by recognizability (mnemonic quality), tradition, and uniqueness when possible.

I am indebted to the literary executor of the late Sir Ronald A. Fisher, F.R.S.; to Dr. Frank Yates, F.R.S.; and to Oliver & Boyd, Edinburgh, for permission to reprint, as Table A.7.1, part of Table 23 from their book, *Statistical Tables for Biological, Agricultural, and Medical Research*. Also, I thank the management of the Rand Corporation (Santa Monica, Calif.), CIBA-GEIGY Ltd. (Basel), Aerospace Research Laboratories of the U.S. Air Force (Dayton, Ohio), Prentice-Hall (Englewood Cliffs, N.J.), Cambridge University Press, and Iowa State University Press (Ames); the editors of the *Annals of Mathematical Statistics, Biometrics, Biometrika, Journal of the American Statistical Association,* and *Technometrics*; and Glenn Beck, A. H. Bowker, H. A. David, C. M. Dayton, C. W. Dunnett, F. E. Grubbs, H. L. Harter, N. A. Hart-

mann, R. B. Howe, D. R. Jensen, G. J. Lieberman, Bernard Ostle, E. S. Pearson, W. D. Schafer, S. S. Shapiro, F. J. Wall, and M. B. Wilk for permission to reproduce various tables or charts published or produced by them. Specific acknowledgments are given as footnotes to the appropriate tables and charts.
        Should errors be found, I would be grateful to hear of them.

DESIGN AND ANALYSIS OF EXPERIMENTS
in the Animal and Medical Sciences

VOLUME 3
APPENDICES

Tables and Figures

TABLE A.1.   TEN THOUSAND RANDOM DIGITS*

　　To select a random digit, one should first find a random starting position. For example, flip a coin and turn to the first two pages of the table
if a head occurs or the last two pages if a tail occurs. Flip again to decide
which of the two pages to use. Then blindly choose a five-digit number on
that page. One may try to avoid the tendency *always* to point to the middle of
the page. Use the first two of the five digits to locate a row in the four
page table and the next two digits to locate a column.
　　The digit appearing at the intersection of the row and column selected is
a random digit in the range 0-9. Or, if one desires a random number in the
range 0-1, the digit at the intersection is the most significant digit of a
decimal fraction and the digits immediately following in the same row may be
used as less significant digits. A random number in the range 0-1 can easily
be converted to a random number in the range 1-n by multiplying the random
number by n, adding 0.5, and rounding the result.
　　For example, suppose the first flip of a coin obtains a head, designating
the first two pages of the table, and the second then obtains a tail, designating the second page of the two. Further, suppose the five-digit number selected blindly is 52166. Then, upon checking the intersection of row 52 and
column 16, one finds the random digit 2, followed by digits 6269, indicating a
random decimal fraction 0.26269. Suppose one wishes to have a random number
in the range 1-24. Then, (24 × 0.26269) + 0.5 = 6.8 or 7. Additional random
numbers may be obtained by proceeding down the column from the position of the
first random number. Of course one must exclude repeated random numbers when
using the table to assign subjects to treatments.

---

　　*Extracted by permission from Rand Corp., 1955, *A million random digits
with 100,000 normal deviates* (New York: Free Press), pp. 1-4.

	00-04	05-09	10-14	15-19	20-24	25-29	30-34	35-39	40-44	45-49
00	10097	32533	76520	13586	34673	54876	80959	09117	39292	74945
01	37542	04805	64894	74296	24805	24037	20636	10402	00822	91665
02	08422	68953	19645	09303	23209	02560	15953	34764	35080	33606
03	99019	02529	09376	70715	38311	31165	88676	74397	04436	27659
04	12807	99970	80157	36147	64032	36653	98951	16877	12171	76833
05	66065	74717	34072	76850	36697	36170	65813	39885	11199	29170
06	31060	10805	45571	82406	35303	42614	86799	07439	23403	09732
07	85269	77602	02051	65692	68665	74818	73053	85247	18623	88579
08	63573	32135	05325	47048	90553	57548	28468	28709	83491	25624
09	73796	45753	03529	64778	35808	34282	60935	20344	35273	88435
10	98520	17767	14905	68607	22109	40558	60970	93433	50500	73998
11	11805	05431	39808	27732	50725	68248	29405	24201	52775	67851
12	83452	99634	06288	98083	13746	70078	18475	40610	68711	77817
13	88685	40200	86507	58401	36766	67951	90364	76493	29609	11062
14	99594	67348	87517	64969	91826	08928	93785	61368	23478	34113
15	65481	17674	17468	50950	58047	76974	73039	57186	40218	16544
16	80124	35635	17727	08015	45318	22374	21115	78253	14385	53763
17	74350	99817	77402	77214	43236	00210	45521	64237	96286	02655
18	69916	26803	66252	29148	36936	87203	76621	13990	94400	56418
19	09893	20505	14225	68514	46427	56788	96297	78822	54382	14598
20	91499	14523	68479	27686	46162	83554	94750	89923	37089	20048
21	80336	94598	26940	36858	70297	34135	53140	33340	42050	82341
22	44104	81949	85157	47954	32979	26575	57600	40881	22222	06413
23	12550	73742	11100	02040	12860	74697	96644	89439	28707	25815
24	63606	49329	16505	34484	40219	52563	43651	77082	07207	31790
25	61196	90446	26457	47774	51924	33729	65394	59593	42582	60527
26	15474	45266	95270	79953	59367	83848	82396	10118	33211	59466
27	94557	28573	67897	54387	54622	44431	91190	42592	92927	45973
28	42481	16213	97344	08721	16868	48767	03071	12059	25701	46670
29	23523	78317	73208	89837	68935	91416	26252	29663	05522	82562
30	04493	52494	75246	33824	45862	51025	61962	79335	65337	12472
31	00549	97654	64051	88159	96119	63896	54692	82391	23287	29529
32	35963	15307	26898	09354	33351	35462	77974	50024	90103	39333
33	59808	08391	45427	26842	83609	49700	13021	24892	78565	20106
34	46058	85236	01390	92286	77281	44077	93910	83647	70617	42941
35	32179	00597	87379	25241	05567	07007	86743	17157	85394	11838
36	69234	61406	20117	45204	15956	60000	18743	92423	97118	96338
37	19565	41430	01758	75379	40419	21585	66674	36806	84962	85207
38	45155	14938	19476	07246	43667	94543	59047	90033	20826	69541
39	94864	31994	36168	10851	34888	81553	01540	35456	05014	51176
40	98086	24826	45240	28404	44999	08896	39094	73407	35441	31880
41	33185	16232	41941	50949	89435	48581	88695	41994	37548	73043
42	80951	00406	96382	70774	20151	23387	25016	25298	94624	61171
43	79752	49140	71961	28296	69861	02591	74852	20539	00387	59579
44	18633	32537	98145	06571	31010	24674	05455	61427	77938	91936
45	74029	43902	77557	32270	97790	17119	52527	58021	80814	51748
46	54178	45611	80993	37143	05335	12969	56127	19255	36040	90324
47	11664	49883	52079	84827	59381	71539	09973	33440	88461	23356
48	48324	77928	31249	64710	02295	36870	32307	57546	15020	09994
49	69074	94138	87637	91976	35584	04401	10518	21615	01848	76938

	00–04	05–09	10–14	15–19	20–24	25–29	30–34	35–39	40–44	45–49
50	09188	20097	32825	39527	04220	86304	83389	87374	64278	58044
51	90045	85497	51981	50654	94938	81997	91870	76150	68476	64659
52	73189	50207	47677	26269	62290	64464	27124	67018	41361	82760
53	75768	76490	20971	87749	90429	12272	95375	05871	93823	43178
54	54016	44056	66281	31003	00682	27398	20714	53295	07706	17813
55	08358	69910	78542	42785	13661	58873	04618	97553	31223	08420
56	28306	03264	81333	10591	40510	07893	32604	60475	94119	01840
57	53840	86233	81594	13628	51215	90290	28466	68795	77762	20791
58	91757	53741	61613	62269	50263	90212	55781	76514	83483	47055
59	89415	92694	00397	58391	12607	17646	48949	72306	94541	37408
60	77513	03820	86864	29901	68414	82774	51908	13980	72893	55507
61	19502	37174	69979	20288	55210	29773	74287	75251	65344	67415
62	21818	59313	93278	81757	05686	73156	07082	85046	31853	38452
63	51474	66499	68107	23621	94049	91345	42836	09191	08007	45449
64	99559	68331	62535	24170	69777	12830	74819	78142	43860	72834
65	33713	48007	93584	72869	51926	64721	58303	29822	93174	93972
66	85274	86893	11303	22970	28834	34137	73515	90400	71148	43643
67	84133	89640	44035	52166	73852	70091	61222	60561	62327	18423
68	56732	16234	17395	96131	10123	91622	85496	57560	81604	18880
69	65138	56806	87648	85261	34313	65861	45875	21069	85644	47277
70	38001	02176	81719	11711	71602	92937	74219	64049	65584	49698
71	37402	96397	01304	77586	56271	10086	47324	62605	40030	37438
72	97125	40348	87083	31417	21815	39250	75237	62047	15501	29578
73	21826	41134	47143	34072	64638	85902	49139	06441	03856	54552
74	73135	42742	95719	09035	85794	74296	08789	88156	64691	19202
75	07638	77929	03061	18072	96207	44156	23821	99538	04713	66994
76	60528	83441	07954	19814	59175	20695	05533	52139	61212	06455
77	83596	35655	06958	92983	05128	09719	77433	53783	92301	50498
78	10850	62746	99599	10507	13499	06319	53075	71839	06410	19362
79	39820	98952	43622	63147	64421	80814	43800	09351	31024	73167
80	59580	06478	75569	78800	88835	54486	23768	06156	04111	08408
81	38508	07341	23793	48763	90822	97022	17719	04207	95954	49953
82	30692	70668	94688	16127	56196	80091	82067	63400	05462	69200
83	65443	95659	18288	27437	49632	24041	08337	65676	96299	90836
84	27267	50264	13192	72294	07477	44606	17985	48911	97341	30358
85	91307	06991	19072	24210	36699	53728	28825	35793	28976	66252
86	68434	94688	84473	13622	62126	98408	12843	82590	09815	93146
87	48908	15877	54745	24591	35700	04754	83824	52692	54130	55160
88	06913	45197	42672	78601	11883	09528	63011	98901	14974	40344
89	10455	16019	14210	33712	91342	37821	88325	80851	43667	70883
90	12883	97343	65027	61184	04285	01392	17974	15077	90712	26769
91	21778	30976	38807	36961	31649	42096	63281	02023	08816	47449
92	19523	59515	65122	59659	86283	68258	69572	13798	16435	91529
93	67245	52670	35583	16563	79246	86686	76463	34222	26655	90802
94	60584	47377	07500	37992	45134	26529	26760	83637	41326	44344
95	53853	41377	36066	94850	58838	73859	49364	73331	96240	43642
96	24637	38736	74384	89342	52623	07992	12369	18601	03742	83873
97	83080	12451	38992	22815	07759	51777	97377	27585	51972	37867
98	16444	24334	36151	99073	27493	70939	85130	32552	54846	54759
99	60790	18157	57178	65762	11161	78576	45819	52979	65130	04860

	50-54	55-59	60-64	65-69	70-74	75-79	80-84	85-89	90-94	95-99
00	03991	10461	93716	16894	66083	24653	84609	58232	88618	19161
01	38555	95554	32886	59780	08355	60860	29735	47762	71299	23853
02	17546	73704	92052	46215	55121	29281	59076	07936	27954	58909
03	32643	52861	95819	06831	00911	98936	76355	93779	80863	00514
04	69572	68777	39510	35905	14060	40619	29549	69616	33564	60780
05	24122	66591	27699	06494	14845	46672	61958	77100	90899	75754
06	61196	30231	92962	61773	41839	55382	17267	70943	78038	70267
07	30532	21704	10274	12202	39685	23309	10061	68829	55986	66485
08	03788	97599	75867	20717	74416	53166	35208	33374	87539	08823
09	48228	63379	85783	47619	53152	67433	35663	52972	16818	60311
10	60365	94653	35075	33949	42614	29297	01918	28316	98953	73231
11	83799	42402	56623	34442	34994	41374	70071	14736	09958	18065
12	32960	07405	36409	83232	99385	41600	11133	07586	15917	06253
13	19322	53845	57620	52606	66497	68646	78138	66559	19640	99413
14	11220	94747	07399	37408	48509	23929	27482	45476	85244	35159
15	31751	57260	68980	05339	15470	48355	88651	22596	03152	19121
16	88492	99382	14454	04504	20094	98977	74873	93413	22109	78508
17	30934	47744	07481	83828	73788	06533	28597	20405	94205	20380
18	22888	48893	27499	98748	60530	45128	74022	84617	82037	10268
19	78212	16993	35902	91386	44372	15486	65741	14014	87481	37220
20	41849	84547	46850	52326	34677	58300	74910	64345	19325	81549
21	46352	33049	69248	93460	45305	07521	61318	31855	14413	70951
22	11087	96294	14013	31792	59747	67277	76503	34513	39663	77544
23	52701	08337	56303	87315	16520	69676	11654	99893	02181	68161
24	57275	36898	81304	48585	68652	27376	92852	55866	88448	03584
25	20857	73156	70284	24326	79375	95220	01159	63267	10622	48391
26	15633	84924	90415	93614	33521	26665	55823	47641	86225	31704
27	92694	48297	39904	02115	59589	49067	66821	41575	49767	04037
28	77613	19019	88152	00080	20554	91409	96277	48257	50816	97616
29	38688	32486	45134	63545	59404	72059	43947	51680	43852	59693
30	25163	01889	70014	15021	41290	67312	71857	15957	68971	11403
31	65251	07629	37239	33295	05870	01119	92784	26340	18477	65622
32	36815	43625	18637	37509	82444	99005	04921	73701	14707	93997
33	64397	11692	05327	82162	20247	81759	45197	25332	83745	22567
34	04515	25624	95096	67946	48460	85558	15191	18782	16930	33361
35	83761	60873	43253	84145	60833	25983	01291	41349	20368	07126
36	14387	06345	80854	09279	43529	06318	38384	74761	41196	37480
37	51321	92246	80088	77074	88722	56736	66164	49431	66919	31678
38	72472	00008	80890	18002	94813	31900	54155	83436	35352	54131
39	05466	55306	93128	18464	74457	90561	72848	11834	79982	68416
40	39528	72484	82474	25593	48545	35247	18619	13674	18611	19241
41	81616	18711	53342	44276	75122	11724	74627	73707	58319	15997
42	07586	16120	82641	22820	92904	13141	32392	19763	61199	67940
43	90767	04235	13574	17200	69902	63742	78464	22501	18627	90872
44	40188	28193	29593	88627	94972	11598	62095	36787	00441	58997
45	34414	82157	86887	55087	19152	00023	12302	80783	32624	68691
46	63439	75363	44989	16822	36024	00867	76378	41605	65961	73488
47	67049	09070	93399	45547	94458	74284	05041	49807	20288	34060
48	79495	04146	52162	90286	54158	34243	46978	35482	59362	95938
49	91704	30552	04737	21031	75051	93029	47665	64382	99782	93478

6

	50-54	55-59	60-64	65-69	70-74	75-79	80-84	85-89	90-94	95-99
50	94015	46874	32444	48277	59820	96163	64654	25843	41145	42820
51	74108	88222	88570	74015	25704	91035	01755	14750	48968	38603
52	62880	87873	95160	59221	22304	90314	72877	17334	39283	04149
53	11748	12102	80580	41867	17710	59621	06554	07850	73950	79552
54	17944	05600	60478	03343	25852	58905	57216	39618	49856	99326
55	66067	42792	95043	52680	46780	56487	09971	59481	37006	22186
56	54244	91030	45547	70818	59849	96169	61459	21647	87417	17198
57	30945	57589	31732	57260	47670	07654	46376	25366	94746	49580
58	69170	37403	86995	90307	94304	71803	26825	05511	12459	91314
59	08345	88975	35841	85771	08105	59987	87112	21476	14713	71181
60	27767	43584	85301	88977	29490	69714	73035	41207	74699	09310
61	13025	14338	54066	15243	47724	66733	47431	43905	31048	56699
62	80217	36292	98525	24335	24432	24896	43277	58874	11466	16082
63	10875	62004	90391	61105	57411	06368	53856	30743	08670	84741
64	54127	57326	26629	19087	24472	88779	30540	27886	61732	75454
65	60311	42824	37301	42678	45990	43242	17374	52003	70707	70214
66	49739	71484	92003	98086	76668	73209	59202	11973	02902	33250
67	78626	51594	16453	94614	39014	97066	83012	09832	25571	77628
68	66692	13986	99837	00582	81232	44987	09504	96412	90193	79568
69	44071	28091	07362	97703	76447	42537	98524	97831	65704	09514
70	41468	85149	49554	17994	14924	39650	95294	00556	70481	06905
71	94559	37559	49678	53119	70312	05682	66986	34099	74474	20740
72	41615	70360	64114	58660	90850	64618	80620	51790	11436	38072
73	50273	93113	41794	86861	24781	89683	55411	85667	77535	99892
74	41396	80504	90670	08289	40902	05069	95083	06783	28102	57816
75	25807	24260	71529	78920	72682	07385	90726	57166	98884	08583
76	06170	97965	88302	98041	21443	41808	68984	83620	89747	98882
77	60808	54444	74412	81105	01176	28838	36421	16489	18059	51061
78	80940	44893	10408	36222	80582	71944	92638	40333	67054	16067
79	19516	90120	46759	71643	13177	55292	21036	82808	77501	97427
80	49386	54480	23604	23554	21785	41101	91178	10174	29420	90438
81	06312	88940	15995	69321	47458	64809	98189	81851	29651	84215
82	60942	00307	11897	92674	40405	68032	96717	54244	10701	41393
83	92329	98932	78284	46347	71209	92061	39448	93136	25722	08564
84	77936	63574	31384	51924	85561	29671	58137	17820	22751	36518
85	38101	77756	11657	13897	95889	57067	47648	13885	70669	93406
86	39641	69457	91339	22502	92613	89719	11947	56203	19324	20504
87	84054	40455	99396	63680	67667	60631	69181	96845	38525	11600
88	47468	03577	57649	63266	24700	71594	14004	23153	69249	05747
89	43321	31370	28977	23896	76479	68562	62342	07589	08899	05985
90	64281	61826	18555	64937	13173	33365	78851	16499	87064	13075
91	66847	70495	32350	02985	87616	38746	26313	77463	55387	72681
92	72461	33230	21529	53424	92581	02262	78438	66276	18396	73538
93	21032	91050	13058	16218	12470	56500	15292	76139	59526	52113
94	95362	67011	06651	16136	01016	00857	55018	56374	35824	71708
95	49712	97380	10404	55452	34030	60726	75211	10271	36633	68424
96	58275	61764	97586	54716	50259	46345	87195	46092	26787	60939
97	89514	11788	68224	23417	73959	76145	30342	40277	11049	72049
98	15472	50669	48139	36732	46874	37088	73465	09819	58869	35220
99	12120	86124	51247	44302	60883	52109	21437	36786	49226	77837

TABLE A.2.   CUMULATIVE STANDARD NORMAL DISTRIBUTION FUNCTION*

The cumulative standard normal distribution function (see Fig. A.2) is

$$F_Z(z_i) = P(Z < z_i) = \alpha \text{ for } z_i < 0 \ (\alpha < 0.5)$$

$$= 1 - \alpha \text{ for } z_i > 0$$

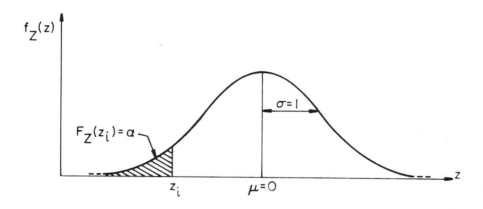

Fig. A.2.   Standard normal distribution curve:
$$F_Z(z_i) = P[Z < z_i] = \alpha \text{ for } z_i < 0 \ (\alpha < 0.5)$$
$$= 1 - \alpha \text{ for } z_i > 0$$

For two-tailed procedures one should seek $z_{\alpha/2}$ and $z_{1-\alpha/2}$ corresponding to the cumulative probabilities $\alpha/2$ and $1-\alpha/2$ in the body of the table. Values commonly used are underlined. However, for more accurate and easy reference, one may use the following chart of percentage points in the upper tail.

$1-\alpha$ or $1-\alpha/2$:	0.8	0.9	0.95	0.975	0.99	0.995	0.999
$z_{1-\alpha}$ or $z_{1-\alpha/2}$:	0.8416	1.2816	1.6449	1.9600	2.3263	2.5758	3.0902

The corresponding points for α or $\alpha/2$ in the lower tail have the same magnitudes but are negative in sign.

---

*Extracted by permission from E. S. Pearson and H. O. Hartley, eds., 1958, *Biometrika tables for statisticians*, Vol. 1, 2d ed. (Cambridge Univ. Press), pp. 104-8.

TABLE A.2. CUMULATIVE STANDARD NORMAL DISTRIBUTION FUNCTION (cont.)

z_i	$F(z_i)$	z_i	$F(z_i)$	z_i	$F(z_i)$	z_i	$F(z_i)$	z_i	$F(z_i)$
-4.00*	.00003	-3.60	.00016	-3.20	.00069	-2.80	.00256	-2.40	.00820
-3.99	.00003	-3.59	.00017	-3.19	.00071	-2.79	.00264	-2.39	.00842
-3.98	.00003	-3.58	.00017	-3.18	.00074	-2.78	.00272	-2.38	.00866
-3.97	.00004	-3.57	.00018	-3.17	.00076	-2.77	.00280	-2.37	.00889
-3.96	.00004	-3.56	.00019	-3.16	.00079	-2.76	.00289	-2.36	.00914
-3.95	.00004	-3.55	.00019	-3.15	.00082	-2.75	.00298	-2.35	.00939
-3.94	.00004	-3.54	.00020	-3.14	.00084	-2.74	.00307	-2.34	.00964
-3.93	.00004	-3.53	.00021	-3.13	.00087	-2.73	.00317	-2.33	.00990
-3.92	.00004	-3.52	.00022	-3.12	.00090	-2.72	.00326	-2.32	.01017
-3.91	.00005	-3.51	.00022	-3.11	.00094	-2.71	.00336	-2.31	.01044
-3.90	.00005	-3.50	.00023	-3.10	.00097	-2.70	.00347	-2.30	.01072
-3.89	.00005	-3.49	.00024	-3.09	.00100	-2.69	.00357	-2.29	.01101
-3.88	.00005	-3.48	.00025	-3.08	.00104	-2.68	.00368	-2.28	.01130
-3.87	.00005	-3.47	.00026	-3.07	.00107	-2.67	.00379	-2.27	.01160
-3.86	.00006	-3.46	.00027	-3.06	.00111	-2.66	.00391	-2.26	.01191
-3.85	.00006	-3.45	.00028	-3.05	.00114	-2.65	.00402	-2.25	.01222
-3.84	.00006	-3.44	.00029	-3.04	.00118	-2.64	.00415	-2.24	.01255
-3.83	.00006	-3.43	.00030	-3.03	.00122	-2.63	.00427	-2.23	.01287
-3.82	.00007	-3.42	.00031	-3.02	.00126	-2.62	.00440	-2.22	.01331
-3.81	.00007	-3.41	.00032	-3.01	.00131	-2.61	.00453	-2.21	.01355
-3.80	.00007	-3.40	.00034	-3.00	.00135	-2.60	.00466	-2.20	.01390
-3.79	.00008	-3.39	.00035	-2.99	.00139	-2.59	.00480	-2.19	.01426
-3.78	.00008	-3.38	.00036	-2.98	.00144	-2.58	.00494	-2.18	.01463
-3.77	.00008	-3.37	.00038	-2.97	.00149	-2.57	.00508	-2.17	.01500
-3.76	.00008	-3.36	.00039	-2.96	.00154	-2.56	.00523	-2.16	.01539
-3.75	.00009	-3.35	.00040	-2.95	.00159	-2.55	.00539	-2.15	.01578
-3.74	.00009	-3.34	.00042	-2.94	.00164	-2.54	.00554	-2.14	.01618
-3.73	.00010	-3.33	.00043	-2.93	.00169	-2.53	.00570	-2.13	.01659
-3.72	.00010	-3.32	.00045	-2.92	.00175	-2.52	.00587	-2.12	.01700
-3.71	.00010	-3.31	.00047	-2.91	.00181	-2.51	.00604	-2.11	.01743
-3.70	.00011	-3.30	.00048	-2.90	.00187	-2.50	.00621	-2.10	.01786
-3.69	.00011	-3.29	.00050	-2.89	.00193	-2.49	.00639	-2.09	.01831
-3.68	.00012	-3.28	.00052	-2.88	.00199	-2.48	.00657	-2.08	.01876
-3.67	.00012	-3.27	.00054	-2.87	.00205	-2.47	.00676	-2.07	.01923
-3.66	.00013	-3.26	.00056	-2.86	.00212	-2.46	.00695	-2.06	.01970
-3.65	.00013	-3.25	.00058	-2.85	.00219	-2.45	.00714	-2.05	.02018
-3.64	.00014	-3.24	.00060	-2.84	.00226	-2.44	.00734	-2.04	.02068
-3.63	.00014	-3.23	.00062	-2.83	.00233	-2.43	.00755	-2.03	.02118
-3.62	.00015	-3.22	.00064	-2.82	.00240	-2.42	.00776	-2.02	.02169
-3.61	.00015	-3.21	.00066	-2.81	.00248	-2.41	.00798	-2.01	.02222

*For $z_i = -5.00$, $F(z_i) = 2.87 \times 10^{-7}$; for $z_i = -6.00$, $F(z_i) = 10^{-10}$.

TABLE A.2. CUMULATIVE STANDARD NORMAL DISTRIBUTION FUNCTION (<u>cont.</u>)

z_i	$F(z_i)$	z_i	$F(z_i)$	z_i	$F(z_i)$	z_i	$F(z_i)$	z_i	$F(z_i)$
-2.00	.02275	-1.60	.05480	-1.20	.11507	-0.80	.21186	-0.40	.34458
-1.99	.02330	-1.59	.05592	-1.19	.11702	-0.79	.21476	-0.39	.34827
-1.98	.02385	-1.58	.05705	-1.18	.11900	-0.78	.21770	-0.38	.35197
-1.97	.02442	-1.57	.05821	-1.17	.12100	-0.77	.22065	-0.37	.35569
-1.96	.02500	-1.56	.05938	-1.16	.12302	-0.76	.22363	-0.36	.35942
-1.95	.02559	-1.55	.06057	-1.15	.12507	-0.75	.22663	-0.35	.36317
-1.94	.02619	-1.54	.06178	-1.14	.12714	-0.74	.22965	-0.34	.36693
-1.93	.02680	-1.53	.06301	-1.13	.12924	-0.73	.23270	-0.33	.37070
-1.92	.02743	-1.52	.06426	-1.12	.13136	-0.72	.23576	-0.32	.37448
-1.91	.02807	-1.51	.06552	-1.11	.13350	-0.71	.23885	-0.31	.37828
-1.90	.02872	-1.50	.06681	-1.10	.13567	-0.70	.24196	-0.30	.38209
-1.89	.02938	-1.49	.06811	-1.09	.13786	-0.69	.24510	-0.29	.38591
-1.88	.03005	-1.48	.06944	-1.08	.14007	-0.68	.24825	-0.28	.38974
-1.87	.03074	-1.47	.07078	-1.07	.14231	-0.67	.25143	-0.27	.39358
-1.86	.03144	-1.46	.07215	-1.06	.14457	-0.66	.25463	-0.26	.39743
-1.85	.03216	-1.45	.07353	-1.05	.14686	-0.65	.25785	-0.25	.40129
-1.84	.03288	-1.44	.07493	-1.04	.14917	-0.64	.26109	-0.24	.40517
-1.83	.03362	-1.43	.07636	-1.03	.15150	-0.63	.26435	-0.23	.40905
-1.82	.03438	-1.42	.07780	-1.02	.15386	-0.62	.26763	-0.22	.41294
-1.81	.03515	-1.41	.07927	-1.01	.15625	-0.61	.27093	-0.21	.41683
-1.80	.03593	-1.40	.08076	-1.00	.15866	-0.60	.27425	-0.20	.42074
-1.79	.03673	-1.39	.08226	-0.99	.16109	-0.59	.27760	-0.19	.42465
-1.78	.03754	-1.38	.08379	-0.98	.16354	-0.58	.28096	-0.18	.42858
-1.77	.03836	-1.37	.08534	-0.97	.16602	-0.57	.28434	-0.17	.43251
-1.76	.03920	-1.36	.08691	-0.96	.16853	-0.56	.28774	-0.16	.43644
-1.75	.04006	-1.35	.08851	-0.95	.17106	-0.55	.29116	-0.15	.44038
-1.74	.04093	-1.34	.09012	-0.94	.17361	-o.54	.29460	-0.14	.44433
-1.73	.04182	-1.33	.09176	-0.93	.17619	-0.53	.29806	-0.13	.44828
-1.72	.04272	-1.32	.09342	-0.92	.17879	-0.52	.30153	-0.12	.45224
-1.71	.04363	-1.31	.09510	-0.91	.18141	-0.51	.30503	-0.11	.45620
-1.70	.04457	-1.30	.09680	-0.90	.18406	-0.50	.30854	-0.10	.46017
-1.69	.04551	-1.29	.09853	-0.89	.18673	-0.49	.31207	-0.09	.46414
-1.68	.04648	-1.28	.10027	-0.88	.18943	-0.48	.31561	-0.08	.46812
-1.67	.04746	-1.27	.10204	-0.87	.19215	-0.47	.31918	-0.07	.47210
-1.66	.04846	-1.26	.10383	-0.86	.19489	-0.46	.32276	-0.06	.47608
-1.65	.04947	-1.25	.10565	-0.85	.19766	-0.45	.32636	-0.05	.48006
-1.64	.05050	-1.24	.10749	-0.84	.20045	-0.44	.32997	-0.04	.48405
-1.63	.05155	-1.23	.10935	-0.83	.20327	-0.43	.33360	-0.03	.48803
-1.62	.05262	-1.22	.11123	-0.82	.20611	-0.42	.33724	-0.02	.49202
-1.61	.05370	-1.21	.11314	-0.81	.20897	-0.41	.34090	-0.01	.49601

TABLE A.2. CUMULATIVE STANDARD NORMAL DISTRIBUTION FUNCTION (cont.)

z_i	$F(z_i)$	z_i	$F(z_i)$	z_i	$F(z_i)$	z_i	$F(z_i)$	z_i	$F(z_i)$
0.01*	.50399	0.41	.65910	0.81	.79103	1.21	.88686	1.61	.94630
0.02	.50798	0.42	.66276	0.82	.79389	1.22	.88877	1.62	.94738
0.03	.51197	0.43	.66640	0.82	.79673	1.23	.89065	1.63	.94845
0.04	.51595	0.44	.67003	0.84	.79955	1.24	.89251	1.64	.94950
0.05	.51994	0.45	.67364	0.85	.80234	1.25	.89435	1.65	.95053
0.06	.52392	0.46	.67724	0.86	.80511	1.26	.89617	1.66	.95154
0.07	.52790	0.47	.68082	0.87	.80785	1.27	.89796	1.67	.95254
0.08	.53188	0.48	.68439	0.88	.81057	1.28	.89973	1.68	.95352
0.09	.53586	0.49	.68793	0.89	.81327	1.29	.90147	1.69	.95449
0.10	.53983	0.50	.69146	0.90	.81594	1.30	.90320	1.70	.95543
0.11	.54380	0.51	.69497	0.91	.81859	1.31	.90490	1.71	.95637
0.12	.54776	0.52	.69847	0.92	.82121	1.32	.90658	1.72	.95728
0.13	.55172	0.53	.70194	0.93	.82381	1.33	.90824	1.73	.95818
0.14	.55567	0.54	.70540	0.94	.82639	1.34	.90988	1.74	.95907
0.15	.55962	0.55	.70884	0.95	.82894	1.35	.91149	1.75	.95994
0.16	.56356	0.56	.71226	0.96	.83147	1.36	.91309	1.76	.96080
0.17	.56749	0.57	.71566	0.97	.83398	1.37	.91466	1.77	.96164
0.18	.57142	0.58	.71904	0.98	.83646	1.38	.91621	1.78	.96246
0.19	.57535	0.59	.72240	0.99	.83891	1.39	.91774	1.79	.96327
0.20	.57926	0.60	.72575	1.00	.84134	1.40	.91924	1.80	.96407
0.21	.58317	0.61	.72907	1.01	.84375	1.41	.92073	1.81	.96485
0.22	.58706	0.62	.73237	1.02	.84614	1.42	.92220	1.82	.96562
0.23	.59095	0.63	.73565	1.03	.84850	1.43	.92364	1.83	.96638
0.24	.59483	0.64	.73891	1.04	.85083	1.44	.92507	1.84	.96712
0.25	.59871	0.65	.74215	1.05	.85314	1.45	.92647	1.85	.96784
0.26	.60257	0.66	.74537	1.06	.85543	1.46	.92785	1.86	.96856
0.27	.60642	0.67	.74857	1.07	.85769	1.47	.92922	1.87	.96926
0.28	.61026	0.68	.75175	1.08	.85993	1.48	.93056	1.88	.96995
0.29	.61409	0.69	.75490	1.09	.86214	1.49	.93189	1.89	.97062
0.30	.61791	0.70	.75804	1.10	.86433	1.50	.93319	1.90	.97128
0.31	.62172	0.71	.76115	1.11	.86650	1.51	.93448	1.91	.97193
0.32	.62552	0.72	.76424	1.12	.86864	1.52	.93574	1.92	.97257
0.33	.62930	0.73	.76730	1.13	.87076	1.53	.93699	1.93	.97320
0.34	.63307	0.74	.77035	1.14	.87286	1.54	.93822	1.94	.97381
0.35	.63683	0.75	.77337	1.15	.87493	1.55	.93943	1.95	.97441
0.36	.64058	0.76	.77637	1.16	.87698	1.56	.94062	1.96	.97500
0.37	.64431	0.77	.77935	1.17	.87900	1.57	.94179	1.97	.97558
0.38	.64803	0.78	.78230	1.18	.88100	1.58	.94295	1.98	.97615
0.39	.65173	0.79	.78524	1.19	.88298	1.59	.94408	1.99	.97670
0.40	.65542	0.80	.78814	1.20	.88493	1.60	.94520	2.00	.97725

*For $z_i = 0.00$, $F(z_i) = 0.50000$.

TABLE A.2.   CUMULATIVE STANDARD NORMAL DISTRIBUTION FUNCTION (<u>cont.</u>)

z_i	$F(z_i)$	z_i	$F(z_i)$	z_i	$F(z_i)$	z_i	$F(z_i)$	z_i	$F(z_i)$
2.01	.97778	2.41	.99202	2.81	.99752	3.21	.99934	3.61	.99985
2.02	.97831	2.42	.99224	2.82	.99760	3.22	.99936	3.62	.99985
2.03	.97882	2.43	.99245	2.83	.99767	3.23	.99938	3.63	.99986
2.04	.97932	2.44	.99266	2.84	.99774	3.24	.99940	3.64	.99986
2.05	.97982	2.45	.99286	2.85	.99781	3.25	.99942	3.65	.99987
2.06	.98030	2.46	.99305	2.86	.99788	3.26	.99944	3.66	.99987
2.07	.98077	2.47	.99324	2.87	.99795	3.27	.99946	3.67	.99988
2.08	.98124	2.48	.99343	2.88	.99801	3.28	.99948	3.68	.99988
2.09	.98169	2.49	.99361	2.89	.99807	3.29	.99950	3.69	.99989
2.10	.98214	2.50	.99379	2.90	.99813	3.30	.99952	3.70	.99989
2.11	.98257	2.51	.99396	2.91	.99819	3.31	.99953	3.71	.99990
2.12	.98300	2.52	.99413	2.92	.99825	3.32	.99955	3.72	.99990
2.13	.98341	2.53	.99430	2.93	.99831	3.33	.99957	3.73	.99990
2.14	.98382	2.54	.99446	2.94	.99836	3.34	.99958	3.74	.99991
2.15	.98422	2.55	.99461	2.95	.99841	3.35	.99960	3.75	.99991
2.16	.98461	2.56	.99477	2.96	.99846	3.36	.99961	3.76	.99992
2.17	.98500	<u>2.57</u>	<u>.99492</u>	2.97	.99851	3.37	.99962	3.77	.99992
2.18	.98537	<u>2.58</u>	<u>.99506</u>	2.98	.99856	3.38	.99964	3.78	.99992
2.19	.98574	2.59	.99520	2.99	.99861	3.39	.99965	3.79	.99992
2.20	.98610	2.60	.99534	3.00	.99865	3.40	.99966	3.80	.99993
2.21	.98645	2.61	.99547	3.01	.99869	3.41	.99968	3.81	.99993
2.22	.98679	2.62	.99560	3.02	.99874	3.42	.99969	3.82	.99993
2.23	.98713	2.63	.99573	3.03	.99878	3.43	.99970	3.83	.99994
2.24	.98745	2.64	.99585	3.04	.99882	3.44	.99971	3.84	.99994
2.25	.98778	2.65	.99598	3.05	.99886	3.45	.99972	3.85	.99994
2.26	.98809	2.66	.99609	3.06	.99889	3.46	.99973	3.86	.99994
2.27	.98840	2.67	.99621	3.07	.99893	3.47	.99974	3.87	.99995
2.28	.98870	2.68	.99632	3.08	.99896	3.48	.99975	3.88	.99995
2.29	.98899	2.69	.99643	3.09	.99900	3.49	.99976	3.89	.99995
2.30	.98928	2.70	.99653	3.10	.99903	3.50	.99977	3.90	.99995
2.31	.98956	2.71	.99664	3.11	.99906	3.51	.99978	3.91	.99995
<u>2.32</u>	<u>.98983</u>	2.72	.99674	3.12	.99910	3.52	.99978	3.92	.99996
<u>2.33</u>	<u>.99010</u>	2.73	.99683	3.13	.99913	3.53	.99979	3.93	.99996
2.34	.99036	2.74	.99693	3.14	.99916	3.54	.99980	3.94	.99996
2.35	.99061	2.75	.99702	3.15	.99918	3.55	.99981	3.95	.99996
2.36	.99086	2.76	.99711	3.16	.99921	3.56	.99981	3.96	.99996
2.37	.99111	2.77	.99720	3.17	.99924	3.57	.99982	3.97	.99996
2.38	.99134	2.78	.99728	3.18	.99926	3.58	.99983	3.89	.99997
2.39	.99158	2.79	.99736	3.19	.99929	3.59	.99983	3.99	.99997
2.40	.99180	2.80	.99744	3.20	.99931	3.60	.99984	4.00*	.99997

*For z_i = 5.00, $F(z_i)$ = 0.9^{6}7133 and for z_i = 6.00, $F(z_i)$ = 0.9^{9}0, where superscripts indicate the number of repetitions of 9 in front of other digits.

TABLE A.3.   PERCENTAGE POINTS OF CHI-SQUARE DISTRIBUTION (1-CDF)

Table A.3.1.   Upper Percentage Points of Chi-Square Distribution

ν	α:	0.3	0.2	0.1	0.05	0.025	0.01	0.005	0.001
1		1.074	1.642	2.706	3.841	5.024	6.635	7.879	10.83
2		2.408	3.219	4.605	5.991	7.378	9.210	10.60	13.82
3		3.665	4.642	6.251	7.815	9.348	11.34	12.84	16.27
4		4.878	5.989	7.779	9.488	11.14	13.28	14.86	18.47
5		6.064	7.289	9.236	11.07	12.83	15.09	16.75	20.52
6		7.231	8.558	10.64	12.59	14.45	16.81	18.55	22.46
7		8.383	9.803	12.02	14.07	16.01	18.48	20.28	24.32
8		9.524	11.03	13.36	15.51	17.53	20.09	21.96	26.12
9		10.66	12.24	14.68	16.92	19.02	21.67	23.59	27.88
10		11.78	13.44	15.99	18.31	20.48	23.21	25.19	29.59
11		12.90	14.63	17.28	19.68	21.92	24.72	26.76	31.26
12		14.01	15.81	18.55	21.03	23.34	26.22	28.30	32.91
13		15.12	16.98	19.81	22.36	24.74	27.69	29.82	34.53
14		16.22	18.15	21.06	23.68	26.12	29.14	31.32	36.12
15		17.32	19.31	22.31	25.00	27.49	30.58	32.80	37.70
16		18.42	20.47	23.54	26.30	28.85	32.00	34.27	39.25
17		19.51	21.61	24.77	27.59	30.19	33.41	35.72	40.79
18		20.60	22.76	25.99	28.87	31.53	34.81	37.16	42.31
19		21.69	23.90	27.20	30.14	32.85	36.19	38.58	43.82
20		22.77	25.04	28.41	31.41	34.17	37.57	40.00	45.31
21		23.86	26.17	29.62	32.67	35.48	38.93	41.40	46.80
22		24.94	27.30	30.81	33.92	36.78	40.29	42.80	48.27
23		26.02	28.43	32.01	35.17	38.08	41.64	44.18	49.73
24		27.10	29.55	33.20	36.42	39.36	42.98	45.56	51.18
25		28.17	30.68	34.38	37.65	40.65	44.31	46.93	52.62
26		29.25	31.79	35.56	38.89	41.92	45.64	48.29	54.05
27		30.32	32.91	36.74	40.11	43.19	46.96	49.64	55.48
28		31.39	34.03	37.92	41.34	44.46	48.28	50.99	56.89
29		32.46	35.14	39.09	42.56	45.72	49.59	52.34	58.30
30		33.53	36.25	40.26	43.77	46.98	50.89	53.67	59.70

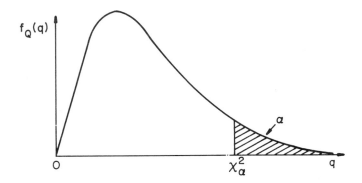

Fig. A.3.1. $P[Q > \chi_\alpha^2] = \alpha$.   For two-tailed procedures, table should be entered at percentage corresponding to $\alpha/2$.

Table A.3.1. Upper Percentage Points of Chi-Square Distribution (<u>cont.</u>)

ν	α: 0.3	0.2	0.1	0.05	0.025	0.01	0.005	0.001
31	34.60	37.36	41.42	44.99	48.23	52.19	55.00	61.10
32	35.66	38.47	42.58	46.19	49.48	53.49	56.33	62.49
33	36.73	39.57	43.75	47.40	50.73	54.78	57.65	63.87
34	37.80	40.68	44.90	48.60	51.97	56.06	58.96	65.25
35	38.86	41.78	46.06	49.80	53.20	57.34	60.27	66.62
36	39.92	42.88	47.21	51.00	54.44	58.62	61.58	67.99
37	40.98	43.98	48.36	52.19	55.67	59.89	62.88	69.35
38	42.05	45.08	49.51	53.38	56.90	61.16	64.18	70.70
39	43.11	46.17	50.66	54.57	58.12	62.43	65.48	72.05
40	44.16	47.27	51.81	55.76	59.34	63.69	66.77	73.40
41	45.22	48.36	52.95	56.94	60.56	64.95	68.05	74.74
42	46.28	49.46	54.09	58.12	61.78	66.21	69.34	76.08
43	47.34	50.55	55.23	59.30	62.99	67.46	70.62	77.42
44	48.40	51.64	56.37	60.48	64.20	68.71	71.89	78.75
45	49.45	52.73	57.51	61.66	65.41	69.96	73.17	80.08
46	50.51	53.82	58.64	62.83	66.62	71.20	74.44	81.40
47	51.56	54.91	59.77	64.00	67.82	72.44	75.70	82.72
48	52.62	55.99	60.91	65.17	69.02	73.68	76.97	84.04
49	53.67	57.08	62.04	66.34	70.22	74.92	78.23	85.35
50	54.72	58.16	63.17	67.50	71.42	76.15	79.49	86.66
51	55.78	59.25	64.30	68.67	72.62	77.39	80.75	87.97
52	56.83	60.33	65.42	69.83	73.81	78.62	82.00	89.27
53	57.88	61.41	66.55	70.99	75.00	79.84	83.25	90.57
54	58.93	62.50	67.67	72.15	76.19	81.07	84.50	91.87
55	59.98	63.58	68.80	73.31	77.38	82.29	85.75	93.17
56	61.03	64.66	69.92	74.47	78.57	83.51	86.99	94.46
57	62.08	65.74	71.04	75.62	79.75	84.73	88.24	95.75
58	63.13	66.82	72.16	76.78	80.94	85.95	89.48	97.04
59	64.18	67.89	73.28	77.93	82.12	87.17	90.72	98.32
60	65.23	68.97	74.40	79.08	83.30	88.38	91.95	99.61
61	66.27	70.05	75.51	80.23	84.48	89.59	93.19	100.9
62	67.32	71.13	76.63	81.38	85.65	90.80	94.42	102.2
63	68.37	72.20	77.75	82.53	86.83	92.01	95.65	103.4
64	69.42	73.28	78.86	83.68	88.00	93.22	96.88	104.7
65	70.46	74.35	79.97	84.82	89.18	94.42	98.11	106.0
66	71.51	75.42	81.09	85.96	90.35	95.63	99.33	107.3
67	72.55	76.50	82.20	87.11	91.52	96.83	100.6	108.5
68	73.60	77.57	83.31	88.25	92.69	98.03	101.8	109.8
69	74.64	78.64	84.42	89.39	93.86	99.23	103.0	111.1
70	75.69	79.71	85.53	90.53	95.02	100.4	104.2	112.3

Table A.3.1. Upper Percentage Points of Chi-Square Distribution (<u>cont.</u>)

ν	α: 0.3	0.2	0.1	0.05	0.025	0.01	0.005	0.001
71	76.73	80.79	86.64	91.67	96.19	101.6	105.4	113.6
72	77.78	81.86	87.74	92.81	97.35	102.8	106.6	114.8
73	78.82	82.93	88.85	93.95	98.52	104.0	107.9	116.1
74	79.86	83.00	89.96	95.08	99.68	105.2	109.1	117.3
75	80.91	85.07	91.06	96.22	100.8	106.4	110.3	118.6
76	81.95	86.13	92.17	97.35	102.0	107.6	111.5	119.8
77	82.99	87.20	93.27	98.48	103.2	108.8	112.7	121.1
78	84.04	88.27	94.37	99.62	104.3	100.0	113.9	122.3
79	85.08	89.34	95.48	100.7	105.5	111.1	115.1	123.6
80	86.12	90.41	96.58	101.9	106.6	112.3	116.3	124.8
81	87.16	91.47	97.68	103.0	107.8	113.5	117.5	126.1
82	88.20	92.54	98.78	104.1	108.9	114.7	118.7	127.3
83	89.24	93.60	99.88	105.3	110.1	115.9	119.9	128.6
84	90.28	94.67	101.0	106.4	111.2	117.1	121.1	129.8
85	91.32	95.73	102.1	107.5	112.4	118.2	122.3	131.0
86	92.36	96.80	103.2	108.6	113.5	119.4	123.5	132.3
87	93.40	97.86	104.3	109.8	114.7	120.6	124.7	133.5
88	94.44	98.93	105.4	110.9	115.8	121.8	125.9	134.7
89	95.48	99.99	106.5	112.0	117.0	122.9	127.1	136.0
90	96.52	101.1	107.6	113.1	118.1	124.1	128.3	137.2
91	97.56	102.1	108.7	114.3	119.3	125.3	129.5	138.4
92	98.60	103.2	109.8	115.4	120.4	126.5	130.7	139.7
93	99.64	104.2	110.8	116.5	121.6	127.6	131.9	140.9
94	100.7	105.3	111.9	117.6	122.7	128.8	133.1	142.1
95	101.7	106.4	113.0	118.8	123.9	130.0	134.2	143.3
96	102.8	107.4	114.1	119.9	125.0	131.1	135.4	144.6
97	103.8	108.5	115.2	121.0	126.1	132.3	136.6	145.8
98	104.8	109.5	116.3	122.1	127.3	133.5	137.8	147.0
99	105.9	110.6	117.4	123.2	128.4	134.6	139.0	148.2
100*	106.9	111.7	118.5	124.3	129.6	135.8	140.2	149.4

Source: Values of $\chi^2_{\alpha,\nu}$ ($\nu = 1,2,\ldots,100$) extracted by permission from H. L. Harter, *Biometrika* 51 (1964):234-39.

*For $\nu > 100$, one may use the approximation $\chi^2_{\alpha,\nu} \approx (z_{1-\alpha} + \sqrt{2\nu-1})^2/2$, where $z_{1-\alpha}$ is an upper percentage point from the standard normal distribution (Table A.2).

Table A.3.2.   Lower Percentage Points of Chi-Square Distribution

ν $1-\alpha$:	0.999	0.995	0.99	0.975	0.95	0.9	0.8	0.7
1	0.002*	0.039*	0.157*	0.982*	0.004	0.016	0.064	0.148
2	0.002	0.010	0.020	0.051	0.103	0.211	0.446	0.713
3	0.024	0.072	0.115	0.216	0.352	0.584	1.005	1.424
4	0.091	0.207	0.297	0.484	0.710	1.064	1.649	2.195
5	0.210	0.412	0.554	0.831	1.145	1.610	2.343	3.000
6	0.381	0.676	0.872	1.237	1.635	2.204	3.070	3.828
7	0.598	0.989	1.239	1.690	2.167	2.833	3.822	4.671
8	0.857	1.344	1.646	2.180	2.733	3.490	4.594	5.527
9	1.152	1.735	2.088	2.700	3.325	4.168	5.380	6.393
10	1.479	2.156	2.558	3.247	3.940	4.865	6.179	7.267
11	1.834	2.603	3.053	3.816	4.575	5.578	6.989	8.148
12	2.214	3.074	3.571	4.404	5.226	6.304	7.807	9.034
13	2.617	3.565	4.107	5.009	5.892	7.042	8.634	9.926
14	3.041	4.075	4.660	5.629	6.571	7.790	9.467	10.82
15	3.483	4.601	5.229	6.262	7.261	8.547	10.31	11.72
16	3.942	5.142	5.812	6.908	7.962	9.312	11.15	12.62
17	4.416	5.697	6.408	7.564	8.672	10.09	12.00	13.53
18	4.905	6.265	7.015	8.231	9.390	10.86	12.86	14.44
19	5.407	6.844	7.633	8.907	10.12	11.65	13.72	15.35
20	5.921	7.434	8.260	9.591	10.85	12.44	14.58	16.27
21	6.447	8.034	8.897	10.28	11.59	13.24	15.44	17.18
22	6.983	8.643	9.542	10.98	12.34	14.04	16.31	18.10
23	7.529	9.260	10.20	11.69	13.09	14.85	17.19	19.02
24	8.085	9.886	10.86	12.40	13.85	15.66	18.06	19.94
25	8.649	10.52	11.52	13.12	14.61	16.47	18.94	20.87
26	9.222	11.16	12.20	13.84	15.38	17.29	19.82	21.79
27	9.803	11.81	12.88	14.57	16.15	18.11	20.70	22.72
28	10.39	12.46	13.56	15.31	16.93	18.94	21.59	23.65
29	10.99	13.12	14.26	16.05	17.71	19.77	22.48	24.58
30	11.59	13.79	14.95	16.79	18.49	20.60	23.36	25.51

*Divide these entries by 1000.

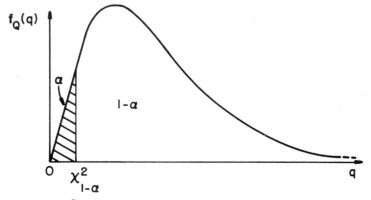

Fig. A.3.2. $P[Q > \chi^2_{1-\alpha}] = 1-\alpha$.   For two-tailed procedures, table should be entered at percentage corresponding to $1-\alpha/2$.

Table A.3.2. Lower Percentage Points of Chi-Square Distribution (<u>cont</u>.)

ν 1-α:	0.999	0.995	0.99	0.975	0.95	0.9	0.8	0.7
31	12.20	14.46	15.66	17.54	19.28	21.43	24.26	26.44
32	12.81	15.13	16.36	18.29	20.07	22.27	25.15	27.37
33	13.43	15.82	17.07	19.05	20.87	23.11	26.04	28.31
34	14.06	16.50	17.79	19.81	21.66	23.95	26.94	29.24
35	14.69	17.19	18.51	20.57	22.46	24.80	27.84	30.18
36	15.32	17.89	19.23	21.34	23.27	25.64	28.74	31.12
37	15.97	18.59	19.96	22.11	24.07	26.49	29.64	32.05
38	16.61	19.29	20.69	22.88	24.88	27.34	30.54	32.99
39	17.26	20.00	21.43	23.65	25.70	28.20	31.44	33.93
40	17.92	20.71	22.16	24.43	26.51	29.05	32.34	34.87
41	18.58	21.42	22.91	25.21	27.33	29.91	33.25	35.81
42	19.24	22.14	23.65	26.00	28.14	30.77	34.16	36.76
43	19.91	22.86	24.40	26.79	28.96	31.63	35.07	37.70
44	20.58	23.58	25.15	27.57	29.79	32.49	35.97	38.64
45	21.25	24.31	25.90	28.37	30.61	33.35	36.88	39.58
46	21.93	25.04	26.66	29.16	31.44	34.22	37.80	40.53
47	22.61	25.77	27.42	29.96	32.27	35.08	38.71	41.47
48	23.29	26.51	28.18	30.75	33.10	35.95	39.62	42.42
49	23.98	27.25	28.94	31.55	33.93	36.82	40.53	43.37
50	24.67	27.99	29.71	32.36	34.76	37.69	41.45	44.31
51	25.37	28.73	30.48	33.16	35.60	38.56	42.36	45.26
52	26.07	29.48	31.25	33.97	36.44	39.43	43.28	46.21
53	26.76	30.23	32.02	34.78	37.28	40.31	44.20	47.16
54	27.47	30.98	32.79	35.59	38.12	41.18	45.12	48.11
55	28.17	31.73	33.57	36.40	38.96	42.06	46.04	49.06
56	28.88	32.49	34.35	37.21	39.80	42.94	46.96	50.01
57	29.59	33.25	35.13	38.03	40.65	43.82	47.88	50.96
58	30.30	34.01	35.91	38.84	41.49	44.70	48.80	51.91
59	31.02	34.77	36.70	39.66	42.34	45.58	49.72	52.86
60	31.74	35.53	37.48	40.48	43.19	46.46	50.64	53.81
61	32.46	36.30	38.27	41.30	44.04	47.34	51.56	54.76
62	33.18	37.07	39.06	42.13	44.89	48.23	52.49	55.71
63	33.91	37.84	39.86	42.95	45.74	49.11	53.41	56.67
64	34.63	38.61	40.65	43.78	46.59	50.00	54.34	57.62
65	35.36	39.38	41.44	44.60	47.45	50.88	55.26	58.57
66	36.09	40.16	42.24	45.43	48.31	51.77	56.19	59.53
67	36.83	40.94	43.04	46.26	49.16	52.66	57.11	60.48
68	37.56	41.71	43.84	47.09	50.02	53.55	58.04	61.44
69	38.30	42.49	44.64	47.92	50.88	54.44	58.97	62.39
70	39.04	43.28	45.44	48.76	51.74	55.33	59.90	63.35

Table A.3.2. Lower Percentage Points of Chi-Square Distribution (<u>cont</u>.)

ν $1-\alpha$:	0.999	0.995	0.99	0.975	0.95	0.9	0.8	0.7
71	39.78	44.06	46.25	49.59	52.60	56.22	60.83	64.30
72	40.52	44.84	47.05	50.43	53.46	57.11	61.76	65.26
73	41.26	45.63	47.86	51.26	54.33	58.01	62.69	66.21
74	42.01	46.42	48.67	52.10	55.19	58.90	63.62	67.17
75	42.76	47.21	49.48	52.94	56.05	59.79	64.55	68.13
76	43.51	48.00	50.29	53.78	56.92	60.69	65.48	69.08
77	44.26	48.79	51.10	54.62	57.79	61.59	66.41	70.04
78	45.01	49.58	51.91	55.47	58.65	62.48	67.34	71.00
79	45.76	50.38	52.72	56.31	59.52	63.38	68.27	71.96
80	46.52	51.17	53.54	57.15	60.39	64.28	69.21	72.92
81	47.28	51.97	54.36	58.00	61.26	65.18	70.14	73.87
82	48.04	52.77	55.17	58.84	62.13	66.08	71.07	74.83
83	48.80	53.57	55.99	59.69	63.00	66.98	72.01	75.79
84	49.56	54.37	56.81	60.54	63.88	67.88	72.94	76.75
85	50.32	55.17	57.63	61.39	64.75	68.78	73.88	77.71
86	51.08	55.97	58.46	62.24	65.62	69.68	74.81	78.67
87	51.85	56.78	59.28	63.09	66.50	70.58	75.75	79.63
88	52.62	57.58	60.10	63.94	67.37	71.48	76.69	80.59
89	53.39	58.39	60.93	64.79	68.25	72.39	77.62	81.55
90	54.16	59.20	61.75	65.65	69.13	73.29	78.56	82.51
91	54.93	60.00	62.58	66.50	70.00	74.20	79.50	83.47
92	55.70	60.81	63.41	67.36	70.88	75.10	80.43	84.43
93	56.47	61.63	64.24	68.21	71.76	76.01	81.37	85.39
94	57.25	62.44	65.07	69.07	72.64	76.91	82.31	86.36
95	58.02	63.25	65.90	69.92	73.52	77.82	83.25	87.32
96	58.80	64.06	66.73	70.78	74.40	78.73	84.19	88.28
97	59.58	64.88	67.56	71.64	75.28	79.63	85.13	89.24
98	60.36	65.69	68.40	72.50	76.16	80.54	86.07	90.20
99	61.14	66.51	69.23	73.36	77.05	81.45	87.01	91.17
100*	61.92	67.33	70.06	74.22	77.93	82.36	87.95	92.13

Source: Values of $\chi^2_{1-\alpha,\nu}$ (ν = 1,2,3,...,100) extracted by permission from H. L. Harter, *Biometrika* 51 (1964):234-39.

   *For $\nu > 100$, one may use the approximation, $\chi^2_{1-\alpha,\nu} = (z_\alpha + \sqrt{2\nu-1})^2/2$, where z_α is a lower percentage point from the standard normal distribution (Table A.2).

TABLE A.4. UPPER PERCENTAGE POINTS OF STUDENT'S t DISTRIBUTION (1-CDF)

ν	0.25	0.20	0.15	0.10	α 0.05	0.025	0.01	0.005	0.0005
1	1.000	1.376	1.963	3.078	6.314	12.706	31.82	63.66	636.6
2	0.816	1.061	1.386	1.886	2.920	4.303	6.965	9.925	31.60
3	0.765	0.978	1.250	1.638	2.353	3.182	4.541	5.841	12.92
4	0.741	0.941	1.190	1.533	2.132	2.776	3.747	4.604	8.610
5	0.727	0.920	1.156	1.476	2.015	2.571	3.365	4.032	6.869
6	0.718	0.906	1.134	1.440	1.943	2.447	3.143	3.707	5.959
7	0.711	0.896	1.119	1.415	1.895	2.365	2.998	3.500	5.408
8	0.706	0.889	1.108	1.397	1.860	2.306	2.896	3.355	5.041
9	0.703	0.883	1.100	1.383	1.833	2.262	2.821	3.250	4.781
10	0.700	0.879	1.093	1.372	1.812	2.228	2.764	3.169	4.587
11	0.698	0.876	1.088	1.363	1.796	2.201	2.718	3.106	4.437
12	0.696	0.873	1.083	1.356	1.782	2.179	2.681	3.054	4.318
13	0.694	0.870	1.079	1.350	1.771	2.160	2.650	3.012	4.221
14	0.692	0.868	1.076	1.345	1.761	2.145	2.624	2.977	4.140
15	0.691	0.866	1.074	1.341	1.753	2.132	2.602	2.947	4.073
16	0.690	0.865	1.071	1.337	1.746	2.120	2.583	2.921	4.015
17	0.689	0.863	1.069	1.333	1.740	2.110	2.567	2.898	3.965
18	0.688	0.862	1.067	1.330	1.734	2.101	2.552	2.878	3.922
19	0.688	0.861	1.066	1.328	1.729	2.093	2.539	2.861	3.883
20	0.687	0.860	1.064	1.325	1.725	2.086	2.528	2.845	3.850
21	0.686	0.859	1.063	1.323	1.721	2.080	2.518	2.831	3.819
22	0.686	0.858	1.061	1.321	1.717	2.074	2.508	2.819	3.792
23	0.685	0.858	1.060	1.319	1.714	2.069	2.500	2.807	3.767
24	0.685	0.857	1.059	1.318	1.711	2.064	2.492	2.797	3.745
25	0.684	0.856	1.058	1.316	1.708	2.060	2.485	2.787	3.725
26	0.684	0.856	1.058	1.315	1.706	2.056	2.479	2.779	3.707
27	0.684	0.855	1.057	1.314	1.703	2.052	2.473	2.771	3.690
28	0.683	0.855	1.056	1.313	1.701	2.048	2.467	2.763	3.674
29	0.683	0.854	1.055	1.311	1.699	2.045	2.462	2.756	3.659
30	0.683	0.854	1.055	1.310	1.697	2.042	2.457	2.750	3.646

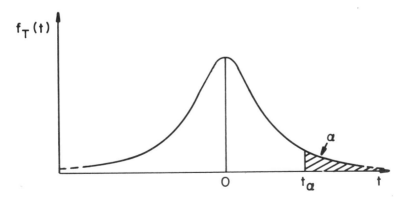

Fig. A.4. $P\left[T > t_\alpha\right] = \alpha$. For two-tailed procedures, table should be entered at column headed by desired value of $\alpha/2$. In all cases, ν = degrees of freedom.

TABLE A.4. UPPER PERCENTAGE POINTS OF STUDENT'S t DISTRIBUTION (1-CDF) (cont.)

ν	0.25	0.20	0.15	0.10	α 0.05	0.025	0.01	0.005	0.0005
31	0.682	0.854	1.054	1.310	1.696	2.040	2.453	2.744	3.634
32	0.682	0.853	1.054	1.309	1.694	2.037	2.449	2.738	3.622
33	0.682	0.853	1.053	1.308	1.692	2.034	2.445	2.733	3.611
34	0.682	0.852	1.053	1.307	1.691	2.032	2.441	2.728	3.601
35	0.682	0.852	1.052	1.306	1.690	2.030	2.438	2.724	3.592
36	0.681	0.852	1.052	1.306	1.688	2.028	2.434	2.720	3.582
37	0.681	0.852	1.051	1.305	1.687	2.026	2.431	2.716	3.574
38	0.681	0.851	1.051	1.304	1.686	2.024	2.428	2.712	3.566
39	0.681	0.851	1.050	1.304	1.685	2.023	2.426	2.708	3.559
40	0.681	0.851	1.050	1.303	1.684	2.021	2.423	2.704	3.551
42	0.680	0.850	1.049	1.302	1.682	2.018	2.418	2.698	3.538
44	0.680	0.850	1.049	1.301	1.680	2.015	2.414	2.692	3.526
46	0.680	0.850	1.048	1.300	1.679	2.013	2.410	2.687	3.515
48	0.680	0.849	1.048	1.299	1.677	2.011	2.406	2.682	3.505
50	0.679	0.849	1.047	1.299	1.676	2.009	2.403	2.678	3.496
60	0.679	0.848	1.046	1.296	1.671	2.000	2.390	2.660	3.461
70	0.678	0.847	1.044	1.294	1.667	1.994	2.381	2.648	3.436
80	0.678	0.846	1.043	1.292	1.664	1.990	2.374	2.639	3.417
90	0.677	0.846	1.042	1.291	1.662	1.987	2.368	2.632	3.402
100	0.677	0.845	1.042	1.290	1.660	1.984	2.364	2.626	3.391
120	0.676	0.845	1.041	1.289	1.658	1.980	2.358	2.618	3.374
140	0.676	0.844	1.040	1.288	1.656	1.977	2.353	2.611	3.362
160	0.676	0.844	1.040	1.287	1.654	1.975	2.350	2.607	3.353
180	0.676	0.844	1.039	1.286	1.653	1.973	2.347	2.604	3.346
200	0.676	0.843	1.039	1.286	1.652	1.972	2.345	2.601	3.340
300	0.676	0.843	1.038	1.285	1.650	1.968	2.338*	2.592	3.323
400	0.676	0.843	1.038	1.284	1.649	1.966	2.335*	2.588	3.315
500	0.676	0.843	1.037	1.284	1.648	1.965	2.334*	2.586	3.310
1000	0.675	0.842	1.037	1.283	1.647	1.962	2.330*	2.581	3.301
∞	0.6745	0.8416	1.0364	1.2816	1.6448	1.9600	2.3263	2.5758	3.2905

Source: Reproduction of values for $\nu \leq 200$ from Documenta Geigy Scientific Tables, 7th ed., by permission of the publishers, CIBA-GEIGY Ltd, Basel, Switzerland. Other values extracted by permission from Bernard Ostle, 1963, *Statistics in research*, 2d ed. (Ames: Iowa State Univ. Press), App. 5.
     *By interpolation.

TABLE A.5.   UPPER PERCENTAGE POINTS OF FISHER'S VARIANCE-RATIO (F) DISTRIBU-
            TION (1-CDF)

   This table is divided into seven sections. In A.5.1, percentage points
for α = 0.05 and α = 0.01 are given. They have been extracted by permission
from tables by B. L. van der Waerden (1969, *Mathematical statistics*, New York:
Springer-Verlag, pp. 340-42), for combinations of degrees of freedom for the
numerator ν_1 = 1(1)20(2)30(10)60(20)100 and degrees of freedom for the denominator ν_2 = 1(1)30(2)50(10)100, 125, 150, 200, 300, 500, 1000. Most of those
values as well as most values for ν_1 = 120 and ∞ and for ν_2 = ∞ were origi-
nally obtained from sources listed below. All other values were obtained by
interpolation or by use of approximations by A. Hald, 1952, *Statistical tables
and formulas* (New York: Wiley), pp. 51-53.
   In A.5.2, percentage points for α = 0.50, with ν_1 = 1(1)9(3)15, 20, 24,
30, 40, 60, 120, ∞ and ν_2 = 1(1)30, 40, 60, 120, ∞ have been extracted by permission from tables by M. Merrington and C. M. Thompson, *Biometrika* 33 (1943):
79.
   In A.5.3-A.5.7, percentage points for α = 0.25, 0.10, 0.025, 0.005 and
0.001, respectively, have been extracted by permission from E. S. Pearson and
H. O. Hartley, eds., 1958, *Biometrika tables for statisticians*, vol. 1, 2d ed.
(Cambridge Univ. Press), pp. 157-63, with corrections by Amos and Pearson,
Biometrika 57 (1970):211. Combinations of ν_1 and ν_2 in each of these tables
are as in Table A.5.2. For two-tailed procedures the tables should be entered
at the percentage equal to half the desired value of α. See Fig. A.5. Lower
percentage points (1-CDF) may be obtained as follows:

$$f_{1-\alpha,\nu_1,\nu_2} = (1/f_{\alpha,\nu_2,\nu_1}) \qquad (\alpha < 0.5)$$

Note the reversal of order of the degrees of freedom ν_1 and ν_2.

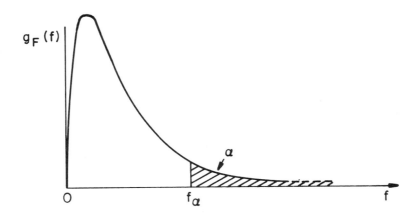

Fig. A.5. $P[F > f_\alpha] = \alpha$. For two-tailed procedures, tables should be entered
           at percentage equal to half the desired value of α.

Hald gave 5% and 1% approximations that may be used in lieu of interpolation for ν_1 and $\nu_2 > 30$:

$$f_{0.05,\nu_1,\nu_2} \approx \{1.4287/\sqrt{[(\nu_1+\nu_2)/2\nu_1\nu_2]-0.95}\}-0.681(\nu_2-\nu_1)/(\nu_1\nu_2)$$

$$f_{0.01,\nu_1,\nu_2} \approx \{2.0206/\sqrt{[(\nu_1+\nu_2)/2\nu_1\nu_2]-1.4}\}-1.073(\nu_2-\nu_1)/(\nu_1\nu_2)$$

More recently, E. E. Johnson, *Technometrics* 15(1973):380, has given quite accurate equations for approximating f values for the upper 5% and 2.5% of the distribution for all combinations of ν_1 and ν_2 up to 120. For $\nu_1 = 1$ and $\nu_2 \geq 5$, the 5% values may be determined from

$$f_{0.05,1,\nu_2} = [7.71-(\nu_2-4.032)/(0.2581\nu_2-0.4076)]$$

For such values the maximum absolute percentage deviation from the nominal 5% is less than 5×10^{-4}%.

For $\nu_1 \geq 2$ and $\nu_2 \geq 5$, the 5% values may be determined from

$$f_{0.05,\nu_1,\nu_2} = [(\nu_1+1.288)/(0.1751\nu_1+0.1129)] - [(\nu_2-4.119)/(0.2511\nu_2-0.4236)]$$

$$- \{0.552-[6.530/(\nu_1+11.533)] - [3.993/(\nu_2+11.533)]\}$$

$$- \{88.889/[(\nu_1+11.533)(\nu_2+11.533)]\}$$

For such values the maximum absolute percentage deviation from the nominal 5% is 0.6%. These equations should be useful in some computer applications in lieu of storage of large tables and interpolation formulas.

Table A.5.1. Upper Percentage Points of F Distribution

ν_2	ν_1:	1	2	3	4	5	6	7	8
1	0.05:	161	200	216	225	230	234	237	239
	0.01:	4052	5000	5403	5625	5764	5859	5928	5982
2	0.05:	18.51	19.0	19.2	19.2	19.3	19.3	19.4	19.4
	0.01:	98.50	99.0	99.2	99.2	99.3	99.3	99.4	99.4
3	0.05:	10.13	9.55	9.28	9.12	9.01	8.94	8.89	8.85
	0.01:	34.12	30.8	29.5	28.7	28.2	27.9	27.7	27.5
4	0.05:	7.71	6.94	6.59	6.39	6.26	6.16	6.09	6.04
	0.01:	21.20	18.0	16.7	16.0	15.5	15.2	15.0	14.8
5	0.05:	6.61	5.79	5.41	5.19	5.05	4.95	4.88	4.82
	0.01:	16.26	13.3	12.1	11.4	11.0	10.7	10.5	10.3
6	0.05:	5.99	5.14	4.76	4.53	4.39	4.28	4.21	4.15
	0.01:	13.75	10.9	9.78	9.15	8.75	8.47	8.26	8.10
7	0.05:	5.59	4.74	4.35	4.12	3.97	3.87	3.79	3.73
	0.01:	12.25	9.55	8.45	7.85	7.46	7.19	6.99	6.84
8	0.05:	5.32	4.46	4.07	3.84	3.69	3.58	3.50	3.44
	0.01:	11.26	8.65	7.59	7.01	6.63	6.37	6.18	6.03
9	0.05:	5.12	4.26	3.86	3.63	3.48	3.37	3.29	3.23
	0.01:	10.56	8.02	6.99	6.42	6.06	5.80	5.61	5.47
10	0.05:	4.96	4.10	3.71	3.48	3.33	3.22	3.14	3.07
	0.01:	10.04	7.56	6.55	5.99	5.64	5.39	5.20	5.06
11	0.05:	4.84	3.98	3.59	3.36	3.20	3.09	3.01	2.95
	0.01:	9.65	7.21	6.22	5.67	5.32	5.07	4.89	4.74
12	0.05:	4.75	3.89	3.49	3.26	3.11	3.00	2.91	2.85
	0.01:	9.33	6.93	5.95	5.41	5.06	4.82	4.64	4.50
13	0.05:	4.67	3.81	3.41	3.18	3.03	2.92	2.83	2.77
	0.01:	9.07	6.70	5.74	5.21	4.86	4.62	4.44	4.30

Table A.5.1. Upper Percentage Points of F Distribution (<u>cont</u>.)

ν_2	ν_1:	1	2	3	4	5	6	7	8
14	0.05:	4.60	3.74	3.34	3.11	2.96	2.85	2.76	2.70
	0.01:	8.86	6.51	5.56	5.04	4.69	4.46	4.28	4.14
15	0.05:	4.54	3.68	3.29	3.06	2.90	2.79	2.71	2.64
	0.01:	8.68	6.36	5.42	4.89	4.56	4.32	4.14	4.00
16	0.05:	4.49	3.63	3.24	3.01	2.85	2.74	2.66	2.59
	0.01:	8.53	6.23	5.29	4.77	4.44	4.20	4.03	3.89
17	0.05:	4.45	3.59	3.20	2.96	2.81	2.70	2.61	2.55
	0.01:	8.40	6.11	5.18	4.67	4.34	4.10	3.93	3.79
18	0.05:	4.41	3.55	3.16	2.93	2.77	2.66	2.58	2.51
	0.01:	8.29	6.01	5.09	4.58	4.25	4.01	3.84	3.71
19	0.05:	4.38	3.52	3.13	2.90	2.74	2.63	2.54	2.48
	0.01:	8.18	5.93	5.01	4.50	4.17	3.94	3.77	3.63
20	0.05:	4.35	3.49	3.10	2.87	2.71	2.60	2.51	2.45
	0.01:	8.10	5.85	4.94	4.43	4.10	3.87	3.70	3.56
21	0.05:	4.32	3.47	3.07	2.84	2.68	2.57	2.49	2.42
	0.01:	8.02	5.78	4.87	4.37	4.04	3.81	3.64	3.51
22	0.05:	4.30	3.44	3.05	2.82	2.66	2.55	2.46	2.40
	0.01:	7.95	5.72	4.82	4.31	3.99	3.76	3.59	3.45
23	0.05:	4.28	3.42	3.03	2.80	2.64	2.53	2.44	2.37
	0.01:	7.88	5.66	4.76	4.26	3.94	3.71	3.54	3.41
24	0.05:	4.26	3.40	3.01	2.78	2.62	2.51	2.42	2.36
	0.01:	7.82	5.61	4.72	4.22	3.90	3.67	3.50	3.36
25	0.05:	4.24	3.39	2.99	2.76	2.60	2.49	2.40	2.34
	0.01:	7.77	5.57	4.68	4.18	3.86	3.63	3.46	3.32
26	0.05:	4.23	3.37	2.98	2.74	2.59	2.47	2.39	2.32
	0.01:	7.72	5.53	4.64	4.14	3.82	3.59	3.42	3.29

Table A.5.1. Upper Percentage Points of F Distribution (<u>cont</u>.)

ν_2	ν_1:	1	2	3	4	5	6	7	8
27	0.05:	4.21	3.35	2.96	2.73	2.57	2.46	2.37	2.31
	0.01:	7.68	5.49	4.60	4.11	3.78	3.56	3.39	3.26
28	0.05:	4.20	3.34	2.95	2.71	2.56	2.45	2.36	2.29
	0.01:	7.64	5.45	4.57	4.07	3.75	3.53	3.36	3.23
29	0.05:	4.18	3.33	2.93	2.70	2.55	2.43	2.35	2.28
	0.01:	7.60	5.42	4.54	4.04	3.73	3.50	3.33	3.20
30	0.05:	4.17	3.32	2.92	2.69	2.53	2.42	2.33	2.27
	0.01:	7.56	5.39	4.51	4.02	3.70	3.47	3.30	3.17
32	0.05:	4.15	3.29	2.90	2.67	2.51	2.40	2.31	2.24
	0.01:	7.50	5.34	4.46	3.97	3.65	3.43	3.26	3.13
34	0.05:	4.13	3.28	2.88	2.65	2.49	2.38	2.29	2.23
	0.01:	7.44	5.29	4.42	3.93	3.61	3.39	3.22	3.09
36	0.05:	4.11	3.26	2.87	2.63	2.48	2.36	2.28	2.21
	0.01:	7.40	5.25	4.38	3.89	3.57	3.35	3.18	3.05
38	0.05:	4.10	3.24	2.85	2.62	2.46	2.35	2.26	2.19
	0.01:	7.35	5.21	4.34	3.86	3.54	3.32	3.15	3.02
40	0.05:	4.08	3.23	2.84	2.61	2.45	2.34	2.25	2.18
	0.01:	7.31	5.18	4.31	3.83	3.51	3.29	3.12	2.99
42	0.05:	4.07	3.22	2.83	2.59	2.44	2.32	2.24	2.17
	0.01:	7.28	5.15	4.29	3.80	3.49	3.27	3.10	2.97
44	0.05:	4.06	3.21	2.82	2.58	2.43	2.31	2.23	2.16
	0.01:	7.25	5.12	4.26	3.78	3.47	3.24	3.08	2.95
46	0.05:	4.05	3.20	2.81	2.57	2.42	2.30	2.22	2.15
	0.01:	7.22	5.10	4.24	3.76	3.44	3.22	3.06	2.93
48	0.05:	4.04	3.19	2.80	2.57	2.41	2.29	2.21	2.14
	0.01:	7.19	5.08	4.22	3.74	3.43	3.20	3.04	2.91

Table A.5.1. Upper Percentage Points of F Distribution (<u>cont</u>.)

ν_2	ν_1:	1	2	3	4	5	6	7	8
50	0.05:	4.03	3.18	2.79	2.56	2.40	2.29	2.20	2.13
	0.01:	7.17	5.06	4.20	3.72	3.41	3.19	3.02	2.89
60	0.05:	4.00	3.15	2.76	2.53	2.37	2.25	2.17	2.10
	0.01:	7.08	4.98	4.13	3.65	3.34	3.12	2.95	2.82
70	0.05:	3.98	3.13	2.74	2.50	2.35	2.23	2.14	2.07
	0.01:	7.01	4.92	4.08	3.60	3.29	3.07	2.91	2.78
80	0.05:	3.96	3.11	2.72	2.49	2.33	2.21	2.13	2.06
	0.01:	6.96	4.88	4.04	3.56	3.26	3.04	2.87	2.74
90	0.05:	3.95	3.10	2.71	2.47	2.32	2.20	2.11	2.04
	0.01:	6.93	4.85	4.01	3.54	3.23	3.01	2.84	2.72
100	0.05:	3.94	3.09	2.70	2.46	2.31	2.19	2.10	2.03
	0.01:	6.90	4.82	3.98	3.51	3.21	2.99	2.82	2.69
125	0.05:	3.92	3.07	2.68	2.44	2.29	2.17	2.08	2.01
	0.01:	6.84	4.78	3.94	3.47	3.17	2.95	2.79	2.66
150	0.05:	3.90	3.06	2.66	2.43	2.27	2.16	2.07	2.00
	0.01:	6.81	4.75	3.92	3.45	3.14	2.92	2.76	2.63
200	0.05:	3.89	3.04	2.65	2.42	2.26	2.14	2.06	1.98
	0.01:	6.76	4.71	3.88	3.41	3.11	2.89	2.73	2.60
300	0.05:	3.87	3.03	2.63	2.40	2.24	2.13	2.04	1.97
	0.01:	6.72	4.68	3.85	3.38	3.08	2.86	2.70	2.57
500	0.05:	3.86	3.01	2.62	2.39	2.23	2.12	2.03	1.96
	0.01:	6.69	4.65	3.82	3.36	3.05	2.84	2.68	2.55
1000	0.05:	3.85	3.00	2.61	2.38	2.22	2.11	2.02	1.95
	0.01:	6.66	4.63	3.80	3.34	3.04	2.82	2.66	2.53
∞	0.05:	3.84	3.00	2.60	2.37	2.21	2.10	2.01	1.94
	0.01:	6.63	4.61	3.78	3.32	3.02	2.80	2.64	2.51

Table A.5.1.   Upper Percentage Points of F Distribution (<u>cont.</u>)

ν_2	ν_1:	9	10	11	12	13	14	15	16
1	0.05:	241	242	243	244	245	245	246	246
	0.01:	6022	6056	6084	6106	6126	6143	6157	6169
2	0.05:	19.4	19.4	19.4	19.4	19.4	19.4	19.4	19.4
	0.01:	99.4	99.4	99.4	99.4	99.4	99.4	99.4	99.4
3	0.05:	8.81	8.79	8.76	8.74	8.73	8.71	8.70	8.69
	0.01:	27.3	27.2	27.1	27.1	27.0	26.9	26.9	26.8
4	0.05:	6.00	5.96	5.94	5.91	5.89	5.87	5.86	5.84
	0.01:	14.7	14.5	14.4	14.4	14.3	14.2	14.2	14.2
5	0.05:	4.77	4.74	4.70	4.68	4.66	4.64	4.62	4.60
	0.01:	10.2	10.1	9.96	9.89	9.82	9.77	9.72	9.68
6	0.05:	4.10	4.06	4.03	4.00	3.98	3.96	3.94	3.92
	0.01:	7.98	7.87	7.79	7.72	7.66	7.60	7.56	7.52
7	0.05:	3.68	3.64	3.60	3.57	3.55	3.53	3.51	3.49
	0.01:	6.72	6.62	6.54	6.47	6.41	6.36	6.31	6.27
8	0.05:	3.39	3.35	3.31	3.28	3.26	3.24	3.22	3.20
	0.01:	5.91	5.81	5.73	5.67	5.61	5.56	5.52	5.48
9	0.05:	3.18	3.14	3.10	3.07	3.05	3.03	3.01	2.99
	0.01:	5.35	5.26	5.18	5.11	5.05	5.00	4.96	4.92
10	0.05:	3.02	2.98	2.94	2.91	2.89	2.86	2.85	2.83
	0.01:	4.94	4.85	4.77	4.71	4.65	4.60	4.56	4.52
11	0.05:	2.90	2.85	2.82	2.79	2.76	2.74	2.72	2.70
	0.01:	4.63	4.54	4.46	4.40	4.34	4.29	4.25	4.21
12	0.05:	2.80	2.75	2.72	2.69	2.66	2.64	2.62	2.60
	0.01:	4.39	4.30	4.22	4.16	4.10	4.05	4.01	3.97
13	0.05:	2.71	2.67	2.63	2.60	2.58	2.55	2.53	2.51
	0.01:	4.19	4.10	4.02	3.96	3.91	3.86	3.82	3.78

Table A.5.1. Upper Percentage Points of F Distribution (<u>cont.</u>)

ν_2	ν_1:	9	10	11	12	13	14	15	16
14	0.05:	2.65	2.60	2.57	2.53	2.51	2.48	2.46	2.44
	0.01:	4.03	3.94	3.86	3.80	3.75	3.70	3.66	3.62
15	0.05:	2.59	2.54	2.51	2.48	2.45	2.42	2.40	2.38
	0.01:	3.89	3.80	3.73	3.67	3.61	3.56	3.52	3.49
16	0.05:	2.54	2.49	2.46	2.42	2.40	2.37	2.35	2.33
	0.01:	3.78	3.69	3.62	3.55	3.50	3.45	3.41	3.37
17	0.05:	2.49	2.45	2.41	2.38	2.35	2.33	2.31	2.29
	0.01:	3.68	3.59	3.52	3.46	3.40	3.35	3.31	3.27
18	0.05:	2.46	2.41	2.37	2.34	2.31	2.29	2.27	2.25
	0.01:	3.60	3.51	3.43	3.37	3.32	3.27	3.23	3.19
19	0.05:	2.42	2.38	2.34	2.31	2.28	2.26	2.23	2.21
	0.01:	3.52	3.43	3.36	3.30	3.24	3.19	3.15	3.12
20	0.05:	2.39	2.35	2.31	2.28	2.25	2.22	2.20	2.18
	0.01:	3.46	3.37	3.29	3.23	3.18	3.13	3.09	3.05
21	0.05:	2.37	2.32	2.28	2.25	2.22	2.20	2.18	2.16
	0.01:	3.40	3.31	3.24	3.17	3.12	3.07	3.03	2.99
22	0.05:	2.34	2.30	2.26	2.23	2.20	2.17	2.15	2.13
	0.01:	3.35	3.26	3.18	3.12	3.07	3.02	2.98	2.94
23	0.05:	2.32	2.27	2.23	2.20	2.18	2.15	2.13	2.11
	0.01:	3.30	3.21	3.14	3.07	3.02	2.97	2.93	2.89
24	0.05:	2.30	2.25	2.21	2.18	2.15	2.13	2.11	2.09
	0.01:	3.26	3.17	3.09	3.03	2.98	2.93	2.89	2.85
25	0.05:	2.28	2.24	2.20	2.16	2.14	2.11	2.09	2.07
	0.01:	3.22	3.13	3.06	2.99	2.94	2.89	2.85	2.81
26	0.05:	2.27	2.22	2.18	2.15	2.12	2.09	2.07	2.05
	0.01:	3.18	3.09	3.02	2.96	2.90	2.86	2.82	2.78

Table A.5.1. Upper Percentage Points of F Distribution (<u>cont</u>.)

ν_2	ν_1:	9	10	11	12	13	14	15	16
27	0.05:	2.25	2.20	2.17	2.13	2.10	2.08	2.06	2.04
	0.01:	3.15	3.06	2.99	2.93	2.87	2.82	2.78	2.75
28	0.05:	2.24	2.19	2.15	2.12	2.09	2.06	2.04	2.02
	0.01:	3.12	3.03	2.96	2.90	2.84	2.79	2.75	2.72
29	0.05:	2.22	2.18	2.14	2.10	2.08	2.05	2.03	2.01
	0.01:	3.09	3.00	2.93	2.87	2.81	2.77	2.73	2.69
30	0.05:	2.21	2.16	2.13	2.09	2.06	2.04	2.01	1.99
	0.01:	3.07	2.98	2.91	2.84	2.79	2.74	2.70	2.66
32	0.05:	2.19	2.14	2.10	2.07	2.04	2.01	1.99	1.97
	0.01:	3.02	2.93	2.86	2.80	2.74	2.70	2.66	2.62
34	0.05:	2.17	2.12	2.08	2.05	2.02	1.99	1.97	1.95
	0.01:	2.98	2.89	2.82	2.76	2.70	2.66	2.62	2.58
36	0.05:	2.15	2.11	2.07	2.03	2.00	1.98	1.95	1.93
	0.01:	2.95	2.86	2.79	2.72	2.67	2.62	2.58	2.54
38	0.05:	2.14	2.09	2.05	2.02	1.99	1.96	1.94	1.92
	0.01:	2.92	2.83	2.75	2.69	2.64	2.59	2.55	2.51
40	0.05:	2.12	2.08	2.04	2.00	1.97	1.95	1.92	1.90
	0.01:	2.89	2.80	2.73	2.66	2.61	2.56	2.52	2.48
42	0.05:	2.11	2.06	2.03	1.99	1.96	1.93	1.91	1.89
	0.01:	2.86	2.78	2.70	2.64	2.59	2.54	2.50	2.46
44	0.05:	2.10	2.05	2.01	1.98	1.95	1.92	1.90	1.88
	0.01:	2.84	2.75	2.68	2.62	2.56	2.52	2.47	2.44
46	0.05:	2.09	2.04	2.00	1.97	1.94	1.91	1.89	1.87
	0.01:	2.82	2.73	2.66	2.60	2.54	2.50	2.45	2.42
48	0.05:	2.08	2.03	1.99	1.96	1.93	1.90	1.88	1.86
	0.01:	2.80	2.72	2.64	2.58	2.53	2.48	2.44	2.40

Table A.5.1. Upper Percentage Points of F Distribution (<u>cont.</u>)

ν_2	ν_1:	9	10	11	12	13	14	15	16
50	0.05:	2.07	2.03	1.99	1.95	1.92	1.89	1.87	1.85
	0.01:	2.79	2.70	2.63	2.56	2.51	2.46	2.42	2.38
60	0.05:	2.04	1.99	1.95	1.92	1.89	1.86	1.84	1.82
	0.01:	2.72	2.63	2.56	2.50	2.44	2.39	2.35	2.31
70	0.05:	2.02	1.97	1.93	1.89	1.86	1.84	1.81	1.79
	0.01:	2.67	2.59	2.51	2.45	2.40	2.35	2.31	2.27
80	0.05:	2.00	1.95	1.91	1.88	1.84	1.82	1.79	1.77
	0.01:	2.64	2.55	2.48	2.42	2.36	2.31	2.27	2.23
90	0.05:	1.99	1.94	1.90	1.86	1.83	1.80	1.78	1.76
	0.01:	2.61	2.52	2.45	2.39	2.33	2.29	2.24	2.21
100	0.05:	1.97	1.93	1.89	1.85	1.82	1.79	1.77	1.75
	0.01:	2.59	2.50	2.43	2.37	2.31	2.26	2.22	2.19
125	0.05:	1.96	1.91	1.87	1.83	1.80	1.77	1.75	1.72
	0.01:	2.55	2.47	2.39	2.33	2.28	2.23	2.19	2.15
150	0.05:	1.94	1.89	1.85	1.82	1.79	1.76	1.73	1.71
	0.01:	2.53	2.44	2.37	2.31	2.25	2.20	2.16	2.12
200	0.05:	1.93	1.88	1.84	1.80	1.77	1.74	1.72	1.69
	0.01:	2.50	2.41	2.34	2.27	2.22	2.17	2.13	2.09
300	0.05:	1.91	1.86	1.82	1.78	1.75	1.72	1.70	1.68
	0.01:	2.47	2.38	2.31	2.24	2.19	2.14	2.10	2.06
500	0.05:	1.90	1.85	1.81	1.77	1.74	1.71	1.69	1.66
	0.01:	2.44	2.36	2.28	2.22	2.17	2.12	2.07	2.04
1000	0.05:	1.89	1.84	1.80	1.76	1.73	1.70	1.68	1.65
	0.01:	2.43	2.34	2.27	2.20	2.15	2.10	2.06	2.02
∞	0.05:	1.88	1.83	1.79	1.75	1.72	1.69	1.67	1.64
	0.01:	2.41	2.32	2.25	2.18	2.13	2.08	2.04	2.00

Table A.5.1. Upper Percentage Points of F Distribution (<u>cont</u>.)

ν_2	ν_1:	17	18	19	20	22	24	26	28
1	0.05:	247	247	248	248	249	249	249	250
	0.01:	6180	6190	6200	6209	6223	6235	6245	6254
2	0.05:	19.4	19.4	19.4	19.4	19.5	19.5	19.5	19.5
	0.01:	99.4	99.4	99.4	99.4	99.5	99.5	99.5	99.5
3	0.05:	8.68	8.67	8.67	8.66	8.65	8.64	8.63	8.62
	0.01:	26.8	26.8	26.7	26.7	26.6	26.6	26.6	26.5
4	0.05:	5.83	5.82	5.81	5.80	5.79	5.77	5.76	5.75
	0.01:	14.1	14.1	14.0	14.0	14.0	13.9	13.9	13.9
5	0.05:	4.59	4.58	4.57	4.56	4.54	4.53	4.52	4.50
	0.01:	9.64	9.61	9.58	9.55	9.51	9.47	9.43	9.40
6	0.05:	3.91	3.90	3.88	3.87	3.86	3.84	3.83	3.82
	0.01:	7.48	7.45	7.42	7.40	7.35	7.31	7.28	7.25
7	0.05:	3.48	3.47	3.46	3.44	3.43	3.41	3.40	3.39
	0.01:	6.24	6.21	6.18	6.16	6.11	6.07	6.04	6.02
8	0.05:	3.19	3.17	3.16	3.15	3.13	3.12	3.10	3.09
	0.01:	5.44	5.41	5.38	5.36	5.32	5.28	5.25	5.22
9	0.05:	2.97	2.96	2.95	2.94	2.92	2.90	2.89	2.87
	0.01:	4.89	4.86	4.83	4.81	4.77	4.73	4.70	4.67
10	0.05:	2.81	2.80	2.78	2.77	2.75	2.74	2.72	2.71
	0.01:	4.49	4.46	4.43	4.41	4.36	4.33	4.30	4.27
11	0.05:	2.69	2.67	2.66	2.65	2.63	2.61	2.59	2.58
	0.01:	4.18	4.15	4.12	4.10	4.06	4.02	3.99	3.96
12	0.05:	2.58	2.57	2.56	2.54	2.52	2.51	2.49	2.48
	0.01:	3.94	3.91	3.88	3.86	3.82	3.78	3.75	3.72
13	0.05:	2.50	2.48	2.47	2.46	2.44	2.42	2.41	2.39
	0.01:	3.75	3.72	3.69	3.66	3.62	3.59	3.56	3.53

Table A.5.1. Upper Percentage Points of F Distribution (<u>cont</u>.)

ν_2	ν_1:	17	18	19	20	22	24	26	28
14	0.05:	2.43	2.41	2.40	2.39	2.37	2.35	2.33	2.32
	0.01:	3.59	3.56	3.53	3.51	3.46	3.43	3.40	3.37
15	0.05:	2.37	2.35	2.34	2.33	2.31	2.29	2.27	2.26
	0.01:	3.45	3.42	3.40	3.37	3.33	3.29	3.26	3.24
16	0.05:	2.32	2.30	2.29	2.28	2.25	2.24	2.22	2.21
	0.01:	3.34	3.31	3.28	3.26	3.22	3.18	3.15	3.12
17	0.05:	2.27	2.26	2.24	2.23	2.21	2.19	2.17	2.16
	0.01:	3.24	3.21	3.18	3.16	3.12	3.08	3.05	3.03
18	0.05:	2.23	2.22	2.20	2.19	2.17	2.15	2.13	2.12
	0.01:	3.16	3.13	3.10	3.08	3.03	3.00	2.97	2.94
19	0.05:	2.20	2.18	2.17	2.16	2.13	2.11	2.10	2.08
	0.01:	3.08	3.05	3.03	3.00	2.96	2.92	2.89	2.87
20	0.05:	2.17	2.15	2.14	2.12	2.10	2.08	2.07	2.05
	0.01:	3.02	2.99	2.96	2.94	2.90	2.86	2.83	2.80
21	0.05:	2.14	2.12	2.11	2.10	2.07	2.05	2.04	2.02
	0.01:	2.96	2.93	2.90	2.88	2.84	2.80	2.77	2.74
22	0.05:	2.11	2.10	2.08	2.07	2.05	2.03	2.01	2.00
	0.01:	2.91	2.88	2.85	2.83	2.78	2.75	2.72	2.69
23	0.05:	2.09	2.07	2.06	2.05	2.02	2.00	1.99	1.97
	0.01:	2.86	2.83	2.80	2.78	2.74	2.70	2.67	2.64
24	0.05:	2.07	2.05	2.04	2.03	2.00	1.98	1.97	1.95
	0.01:	2.82	2.79	2.76	2.74	2.70	2.66	2.63	2.60
25	0.05:	2.05	2.04	2.02	2.01	1.98	1.96	1.95	1.93
	0.01:	2.78	2.75	2.72	2.70	2.66	2.62	2.59	2.56
26	0.05:	2.03	2.02	2.00	1.99	1.97	1.95	1.93	1.91
	0.01:	2.74	2.72	2.69	2.66	2.62	2.58	2.55	2.53

Table A.5.1.   Upper Percentage Points of F Distribution (<u>cont.</u>)

ν_2	ν_1:	17	18	19	20	22	24	26	28
27	0.05:	2.02	2.00	1.99	1.97	1.95	1.93	1.91	1.90
	0.01:	2.71	2.68	2.66	2.63	2.59	2.55	2.52	2.49
28	0.05:	2.00	1.99	1.97	1.96	1.93	1.91	1.90	1.88
	0.01:	2.68	2.65	2.63	2.60	2.56	2.52	2.49	2.46
29	0.05:	1.99	1.97	1.96	1.94	1.92	1.90	1.88	1.87
	0.01:	2.66	2.63	2.60	2.57	2.53	2.49	2.46	2.44
30	0.05:	1.98	1.96	1.95	1.93	1.91	1.89	1.87	1.85
	0.01:	2.63	2.60	2.57	2.55	2.51	2.47	2.44	2.41
32	0.05:	1.95	1.94	1.92	1.91	1.88	1.86	1.85	1.83
	0.01:	2.58	2.55	2.53	2.50	2.46	2.42	2.39	2.36
34	0.05:	1.93	1.92	1.90	1.89	1.86	1.84	1.82	1.80
	0.01:	2.55	2.51	2.49	2.46	2.42	2.38	2.35	2.32
36	0.05:	1.92	1.90	1.88	1.87	1.85	1.82	1.81	1.79
	0.01:	2.51	2.48	2.45	2.43	2.38	2.35	2.32	2.29
38	0.05:	1.90	1.88	1.87	1.85	1.83	1.81	1.79	1.77
	0.01:	2.48	2.45	2.42	2.40	2.35	2.32	2.28	2.26
40	0.05:	1.89	1.87	1.85	1.84	1.81	1.79	1.77	1.76
	0.01:	2.45	2.42	2.39	2.37	2.33	2.29	2.26	2.23
42	0.05:	1.87	1.86	1.84	1.83	1.80	1.78	1.76	1.74
	0.01:	2.43	2.40	2.37	2.34	2.30	2.26	2.23	2.20
44	0.05:	1.86	1.84	1.83	1.81	1.79	1.77	1.75	1.73
	0.01:	2.40	2.37	2.35	2.32	2.28	2.24	2.21	2.18
46	0.05:	1.85	1.83	1.82	1.80	1.78	1.76	1.74	1.72
	0.01:	2.38	2.35	2.33	2.30	2.26	2.22	2.19	2.16
48	0.05:	1.84	1.82	1.81	1.79	1.77	1.75	1.73	1.71
	0.01:	2.37	2.33	2.31	2.28	2.24	2.20	2.17	2.14

Table A.5.1. Upper Percentage Points of F Distribution (<u>cont</u>.)

ν_2	ν_1:	17	18	19	20	22	24	26	28
50	0.05:	1.83	1.81	1.80	1.78	1.76	1.74	1.72	1.70
	0.01:	2.35	2.32	2.29	2.27	2.22	2.18	2.15	2.12
60	0.05:	1.80	1.78	1.76	1.75	1.72	1.70	1.68	1.66
	0.01:	2.28	2.25	2.22	2.20	2.15	2.12	2.08	2.05
70	0.05:	1.77	1.75	1.74	1.72	1.70	1.67	1.65	1.64
	0.01:	2.23	2.20	2.18	2.15	2.11	2.07	2.03	2.01
80	0.05:	1.75	1.73	1.72	1.70	1.68	1.65	1.63	1.62
	0.01:	2.20	2.17	2.14	2.12	2.07	2.03	2.00	1.97
90	0.05:	1.74	1.72	1.70	1.69	1.66	1.64	1.62	1.60
	0.01:	2.17	2.14	2.11	2.09	2.04	2.00	1.97	1.94
100	0.05:	1.73	1.71	1.69	1.68	1.65	1.63	1.61	1.59
	0.01:	2.15	2.12	2.09	2.07	2.02	1.98	1.94	1.92
125	0.05:	1.70	1.69	1.67	1.65	1.63	1.60	1.58	1.57
	0.01:	2.11	2.08	2.05	2.03	1.98	1.94	1.91	1.88
150	0.05:	1.69	1.67	1.66	1.64	1.61	1.59	1.57	1.55
	0.01:	2.09	2.06	2.03	2.00	1.96	1.92	1.88	1.85
200	0.05:	1.67	1.66	1.64	1.62	1.60	1.57	1.55	1.53
	0.01:	2.06	2.02	2.00	1.97	1.93	1.89	1.85	1.82
300	0.05:	1.66	1.64	1.62	1.61	1.58	1.55	1.53	1.51
	0.01:	2.03	1.99	1.97	1.94	1.89	1.85	1.82	1.79
500	0.05:	1.64	1.62	1.61	1.59	1.56	1.54	1.52	1.50
	0.01:	2.00	1.97	1.94	1.92	1.87	1.83	1.79	1.76
1000	0.05:	1.63	1.61	1.60	1.58	1.55	1.53	1.51	1.49
	0.01:	1.98	1.95	1.92	1.90	1.85	1.81	1.77	1.74
∞	0.05:	1.62	1.60	1.59	1.57	1.54	1.52	1.50	1.48
	0.01:	1.96	1.93	1.90	1.88	1.83	1.79	1.75	1.72

Table A.5.1. Upper Percentage Points of F Distribution (<u>cont</u>.)

ν_2	ν_1:	30	40	50	60	80	100	120	∞
1	0.05:	250	251	252	252	252	253	253	254
	0.01:	6261	6287	6302	6313	6323	6332	6339	6366
2	0.05:	19.5	19.5	19.5	19.5	19.5	19.5	19.5	19.5
	0.01:	99.5	99.5	99.5	99.5	99.5	99.5	99.5	99.5
3	0.05:	8.62	8.59	8.58	8.57	8.56	8.55	8.55	8.53
	0.01:	26.5	26.4	26.4	26.3	26.3	26.2	26.2	26.1
4	0.05:	5.75	5.72	5.70	5.69	5.67	5.66	5.66	5.63
	0.01:	13.8	13.7	13.7	13.7	13.6	13.6	13.6	13.5
5	0.05:	4.50	4.46	4.44	4.43	4.41	4.41	4.40	4.36
	0.01:	9.38	9.29	9.24	9.20	9.16	9.13	9.11	9.02
6	0.05:	3.81	3.77	3.75	3.74	3.72	3.71	3.70	3.67
	0.01:	7.23	7.14	7.09	7.06	7.01	6.99	6.97	6.88
7	0.05:	3.38	3.34	3.32	3.30	3.29	3.27	3.27	3.23
	0.01:	5.99	5.91	5.86	5.82	5.78	5.75	5.74	5.65
8	0.05:	3.08	3.04	3.02	3.01	2.99	2.97	2.97	2.93
	0.01:	5.20	5.12	5.07	5.03	4.99	4.96	4.95	4.86
9	0.05:	2.86	2.83	2.80	2.79	2.77	2.76	2.75	2.71
	0.01:	4.65	4.57	4.52	4.48	4.44	4.42	4.40	4.31
10	0.05:	2.70	2.66	2.64	2.62	2.60	2.59	2.58	2.54
	0.01:	4.25	4.17	4.12	4.08	4.04	4.01	4.00	3.91
11	0.05:	2.57	2.53	2.51	2.49	2.47	2.46	2.45	2.40
	0.0i:	3.94	3.86	3.81	3.78	3.73	3.71	3.69	3.60
12	0.05:	2.47	2.43	2.40	2.38	2.36	2.35	2.34	2.30
	0.01:	3.70	3.62	3.57	3.54	3.49	3.47	3.45	3.36
13	0.05:	2.38	2.34	2.31	2.30	2.27	2.26	2.25	2.21
	0.01:	3.51	3.43	3.38	3.34	3.30	3.27	3.25	3.17

Table A.5.1. Upper Percentage Points of F Distribution (<u>cont</u>.)

ν_2	ν_1:	30	40	50	60	80	100	120	∞
14	0.05:	2.31	2.27	2.24	2.22	2.20	2.19	2.18	2.13
	0.01:	3.35	3.27	3.22	3.18	3.14	3.11	3.09	3.00
15	0.05:	2.25	2.20	2.18	2.16	2.14	2.12	2.11	2.07
	0.01:	3.21	3.13	3.08	3.05	3.00	2.98	2.96	2.87
16	0.05:	2.19	2.15	2.12	2.11	2.08	2.07	2.06	2.01
	0.01:	3.10	3.02	2.97	2.93	2.89	2.86	2.84	2.75
17	0.05:	2.15	2.10	2.08	2.06	2.03	2.02	2.01	1.96
	0.01:	3.00	2.92	2.87	2.83	2.79	2.76	2.75	2.65
18	0.05:	2.11	2.06	2.04	2.02	1.99	1.98	1.97	1.92
	0.01:	2.92	2.84	2.78	2.75	2.70	2.68	2.66	2.57
19	0.05:	2.07	2.03	2.00	1.98	1.96	1.94	1.93	1.88
	0.01:	2.84	2.76	2.71	2.67	2.63	2.60	2.58	2.49
20	0.05:	2.04	1.99	1.97	1.95	1.92	1.91	1.90	1.84
	0.01:	2.78	2.69	2.64	2.61	2.56	2.54	2.52	2.42
21	0.05:	2.01	1.96	1.94	1.92	1.89	1.88	1.87	1.81
	0.01:	2.72	2.64	2.58	2.55	2.50	2.48	2.46	2.36
22	0.05:	1.98	1.94	1.91	1.89	1.86	1.85	1.84	1.78
	0.01:	2.67	2.58	2.53	2.50	2.45	2.42	2.40	2.31
23	0.05:	1.96	1.91	1.88	1.86	1.84	1.82	1.81	1.76
	0.01:	2.62	2.54	2.48	2.45	2.40	2.37	2.35	2.26
24	0.05:	1.94	1.89	1.86	1.84	1.82	1.80	1.79	1.73
	0.01:	2.58	2.49	2.44	2.40	2.36	2.33	2.31	2.21
25	0.05:	1.92	1.87	1.84	1.82	1.80	1.78	1.77	1.71
	0.01:	2.54	2.45	2.40	2.36	2.32	2.29	2.27	2.17
26	0.05:	1.90	1.85	1.82	1.80	1.78	1.76	1.75	1.69
	0.01:	2.50	2.42	2.36	2.33	2.28	2.25	2.23	2.13

Table A.5.1. Upper Percentage Points of F Distribution (<u>cont.</u>)

ν_2	ν_1:	30	40	50	60	80	100	120	∞
27	0.05:	1.88	1.84	1.81	1.79	1.76	1.74	1.73	1.67
	0.01:	2.47	2.38	2.33	2.29	2.25	2.22	2.20	2.10
28	0.05:	1.87	1.82	1.79	1.77	1.74	1.73	1.71	1.65
	0.01:	2.44	2.35	2.30	2.26	2.22	2.19	2.17	2.06
29	0.05:	1.85	1.81	1.77	1.75	1.73	1.71	1.70	1.64
	0.01:	2.41	2.33	2.27	2.23	2.19	2.16	2.14	2.03
30	0.05:	1.84	1.79	1.76	1.74	1.71	1.70	1.68	1.62
	0.01:	2.39	2.30	2.25	2.21	2.16	2.13	2.11	2.01
32	0.05:	1.82	1.77	1.74	1.71	1.69	1.67	1.65	1.59
	0.01:	2.34	2.25	2.20	2.16	2.11	2.08	2.06	1.96
34	0.05:	1.80	1.75	1.71	1.69	1.66	1.65	1.63	1.57
	0.01:	2.30	2.21	2.16	2.12	2.07	2.04	2.02	1.91
36	0.05:	1.78	1.73	1.69	1.67	1.64	1.62	1.61	1.55
	0.01:	2.26	2.17	2.12	2.08	2.03	2.00	1.98	1.87
38	0.05:	1.76	1.71	1.68	1.65	1.62	1.61	1.59	1.53
	0.01:	2.23	2.14	2.09	2.05	2.00	1.97	1.95	1.84
40	0.05:	1.74	1.69	1.66	1.64	1.61	1.59	1.58	1.51
	0.01:	2.20	2.11	2.06	2.02	1.97	1.94	1.92	1.81
42	0.05:	1.73	1.68	1.65	1.62	1.59	1.57	1.56	1.49
	0.01:	2.18	2.09	2.03	1.99	1.94	1.91	1.89	1.78
44	0.05:	1.72	1.67	1.63	1.61	1.58	1.56	1.55	1.48
	0.01:	2.15	2.06	2.01	1.97	1.92	1.89	1.87	1.75
46	0.05:	1.71	1.65	1.62	1.60	1.57	1.55	1.54	1.46
	0.01:	2.13	2.04	1.99	1.95	1.90	1.86	1.84	1.72
48	0.05:	1.70	1.64	1.61	1.59	1.56	1.54	1.53	1.45
	0.01:	2.12	2.02	1.97	1.93	1.88	1.84	1.82	1.70

Table A.5.1. Upper Percentage Points of F Distribution (<u>cont</u>.)

ν_2	ν_1:	30	40	50	60	80	100	120	∞
50	0.05:	1.69	1.63	1.60	1.58	1.54	1.52	1.51	1.44
	0.01:	2.10	2.01	1.95	1.91	1.86	1.82	1.80	1.68
60	0.05:	1.65	1.59	1.56	1.53	1.50	1.48	1.47	1.39
	0.01:	2.03	1.94	1.88	1.84	1.78	1.75	1.73	1.60
70	0.05:	1.62	1.57	1.53	1.50	1.47	1.45	1.44	1.35
	0.01:	1.98	1.89	1.83	1.78	1.73	1.70	1.67	1.53
80	0.05:	1.60	1.54	1.51	1.48	1.45	1.43	1.42	1.32
	0.01:	1.94	1.85	1.79	1.75	1.69	1.66	1.63	1.49
90	0.05:	1.59	1.53	1.49	1.46	1.43	1.41	1.40	1.30
	0.01:	1.92	1.82	1.76	1.72	1.66	1.62	1.60	1.46
100	0.05:	1.57	1.52	1.48	1.45	1.41	1.39	1.38	1.28
	0.01:	1.89	1.80	1.73	1.69	1.63	1.60	1.57	1.43
125	0.05:	1.55	1.49	1.45	1.42	1.39	1.36	1.35	1.25
	0.01:	1.85	1.76	1.69	1.65	1.59	1.55	1.53	1.37
150	0.05:	1.53	1.48	1.44	1.41	1.37	1.34	1.33	1.22
	0.01:	1.83	1.73	1.66	1.62	1.56	1.52	1.50	1.33
200	0.05:	1.52	1.46	1.41	1.39	1.35	1.32	1.31	1.19
	0.01:	1.79	1.69	1.63	1.58	1.52	1.48	1.46	1.28
300	0.05:	1.50	1.43	1.39	1.36	1.32	1.30	1.29	1.15
	0.01:	1.76	1.66	1.59	1.55	1.48	1.44	1.42	1.22
500	0.05:	1.48	1.42	1.38	1.34	1.30	1.28	1.27	1.11
	0.01:	1.74	1.63	1.56	1.52	1.45	1.41	1.38	1.16
1000	0.05:	1.47	1.41	1.36	1.33	1.29	1.26	1.24	1.08
	0.01:	1.72	1.61	1.54	1.50	1.43	1.38	1.35	1.11
∞	0.05:	1.46	1.39	1.35	1.32	1.27	1.24	1.22	1.00
	0.01:	1.70	1.59	1.52	1.47	1.40	1.36	1.32	1.00

Source: Extracted by permission from tables by Van der Waerden or de-rived by interpolation or Hald's approximations (see notes preceding Table A.5.1).

Table A.5.2. Upper Percentage Points of F Distribution (α = 0.50)

ν_2 \ ν_1:	1	2	3	4	5	6	7	8	9
1	1.00	1.50	1.71	1.82	1.89	1.94	1.98	2.00	2.03
2	0.67	1.00	1.13	1.21	1.25	1.28	1.30	1.32	1.33
3	0.58	0.88	1.00	1.06	1.10	1.13	1.15	1.16	1.17
4	0.55	0.83	0.94	1.00	1.04	1.06	1.08	1.09	1.10
5	0.53	0.80	0.91	0.96	1.00	1.02	1.04	1.05	1.06
6	0.52	0.78	0.89	0.94	0.98	1.00	1.02	1.03	1.04
7	0.51	0.77	0.87	0.93	0.96	0.98	1.00	1.01	1.02
8	0.50	0.76	0.86	0.92	0.95	0.97	0.99	1.00	1.01
9	0.49	0.75	0.85	0.91	0.94	0.96	0.98	0.99	1.00
10	0.49	0.74	0.84	0.90	0.93	0.95	0.97	0.98	0.99
11	.486	.739	.840	.893	.926	.948	.964	.977	.986
12	.484	.735	.835	.888	.921	.943	.959	.972	.981
13	.481	.731	.832	.885	.917	.939	.955	.967	.977
14	.479	.729	.828	.881	.914	.936	.952	.964	.973
15	.478	.726	.826	.878	.911	.933	.948	.960	.970
16	.476	.724	.823	.876	.908	.930	.946	.958	.967
17	.475	.722	.821	.874	.906	.928	.943	.955	.965
18	.474	.721	.819	.872	.904	.926	.941	.953	.962
19	.473	.719	.818	.870	.902	.924	.939	.951	.961
20	.472	.718	.816	.868	.900	.922	.938	.950	.959
21	.471	.717	.815	.867	.899	.921	.936	.948	.957
22	.470	.715	.814	.866	.898	.919	.935	.947	.956
23	.470	.714	.813	.864	.896	.918	.934	.945	.955
24	.469	.714	.812	.863	.895	.917	.932	.944	.953
25	.468	.713	.811	.862	.894	.916	.931	.943	.952
26	.468	.712	.810	.861	.893	.915	.930	.942	.951
27	.467	.711	.809	.861	.892	.914	.930	.941	.950
28	.467	.711	.808	.860	.892	.913	.929	.940	.950
29	.467	.710	.808	.859	.891	.912	.928	.940	.949
30	.466	.709	.807	.858	.890	.912	.927	.939	.948
40	.463	.705	.802	.854	.885	.907	.922	.934	.943
60	.461	.701	.798	.849	.880	.901	.917	.928	.937
120	.458	.697	.793	.844	.875	.896	.912	.923	.932
∞	.455	.693	.789	.839	.870	.891	.907	.918	.927

Table A.5.2. Upper Percentage Points of F Distribution (α = 0.50) (<u>cont</u>.)

ν_2 \ ν_1:	12	15	20	24	30	40	60	120	∞
1	2.07	2.09	2.12	2.13	2.15	2.16	2.17	2.18	2.20
2	1.36	1.38	1.39	1.40	1.41	1.42	1.43	1.43	1.44
3	1.20	1.21	1.23	1.23	1.24	1.25	1.25	1.26	1.27
4	1.13	1.14	1.15	1.16	1.16	1.17	1.18	1.18	1.19
5	1.09	1.10	1.11	1.12	1.12	1.13	1.14	1.14	1.15
6	1.06	1.07	1.08	1.09	1.10	1.10	1.11	1.12	1.12
7	1.04	1.05	1.07	1.07	1.08	1.08	1.09	1.10	1.10
8	1.03	1.04	1.05	1.06	1.07	1.08	1.08	1.08	1.09
9	1.02	1.03	1.04	1.05	1.05	1.06	1.07	1.07	1.08
10	1.01	1.02	1.03	1.04	1.05	1.05	1.06	1.06	1.07
11	1.01	1.02	1.03	1.03	1.04	1.05	1.05	1.06	1.06
12	1.00	1.01	1.02	1.03	1.03	1.04	1.05	1.05	1.06
13	1.00	1.01	1.02	1.02	1.03	1.04	1.04	1.05	1.05
14	0.99	1.00	1.01	1.02	1.03	1.03	1.04	1.04	1.05
15	0.99	1.00	1.01	1.02	1.02	1.03	1.03	1.04	1.05
16	0.99	1.00	1.01	1.01	1.02	1.03	1.03	1.04	1.04
17	0.98	1.00	1.01	1.01	1.02	1.02	1.03	1.03	1.04
18	0.98	0.99	1.00	1.01	1.02	1.02	1.03	1.03	1.04
19	0.98	0.99	1.00	1.01	1.01	1.02	1.02	1.03	1.04
20	0.98	0.99	1.00	1.01	1.01	1.02	1.02	1.03	1.03
21	0.98	0.99	1.00	1.00	1.01	1.02	1.02	1.03	1.03
22	0.97	0.99	1.00	1.00	1.01	1.01	1.02	1.03	1.03
23	0.97	0.98	1.00	1.00	1.01	1.01	1.02	1.02	1.03
24	0.97	0.98	0.99	1.00	1.01	1.01	1.02	1.02	1.03
25	0.97	0.98	0.99	1.00	1.00	1.01	1.02	1.02	1.03
26	0.97	0.98	0.99	1.00	1.00	1.01	1.01	1.02	1.03
27	0.97	0.98	0.99	1.00	1.00	1.01	1.01	1.02	1.03
28	0.97	0.98	0.99	1.00	1.00	1.01	1.01	1.02	1.02
29	0.97	0.98	0.99	1.00	1.00	1.01	1.01	1.02	1.02
30	0.97	0.98	0.99	0.99	1.00	1.01	1.01	1.02	1.02
40	.961	.972	.983	.989	.994	1.00	1.01	1.01	1.02
60	.956	.967	.978	.983	.989	.994	1.00	1.01	1.01
120	.950	.961	.972	.978	.983	.989	.994	1.00	1.01
∞	.945	.956	.967	.972	.978	.983	.989	.994	1.00

Source: Extracted by permission from tables by Merrington and Thompson (see notes preceding Table A.5.1).

Table A.5.3.   Upper Percentage Points of F Distribution ($\alpha = 0.25$)

ν_2 ν_1:	1	2	3	4	5	6	7	8	9
1	5.83	7.50	8.20	8.58	8.82	8.98	9.10	9.19	9.26
2	2.57	3.00	3.15	3.23	3.28	3.31	3.34	3.35	3.37
3	2.02	2.28	2.36	2.39	2.41	2.42	2.43	2.44	2.44
4	1.81	2.00	2.05	2.06	2.07	2.08	2.08	2.08	2.08
5	1.69	1.85	1.88	1.89	1.89	1.89	1.89	1.89	1.89
6	1.62	1.76	1.78	1.79	1.79	1.78	1.78	1.78	1.77
7	1.57	1.70	1.72	1.72	1.71	1.71	1.70	1.70	1.69
8	1.54	1.66	1.67	1.66	1.66	1.65	1.64	1.64	1.63
9	1.51	1.62	1.63	1.63	1.62	1.61	1.60	1.60	1.59
10	1.49	1.60	1.60	1.59	1.59	1.58	1.57	1.56	1.56
11	1.47	1.58	1.58	1.57	1.56	1.55	1.54	1.53	1.53
12	1.46	1.56	1.56	1.55	1.54	1.53	1.52	1.51	1.51
13	1.45	1.55	1.55	1.53	1.52	1.51	1.50	1.49	1.49
14	1.44	1.53	1.53	1.52	1.51	1.50	1.49	1.48	1.47
15	1.43	1.52	1.52	1.51	1.49	1.48	1.47	1.46	1.46
16	1.42	1.51	1.51	1.50	1.48	1.47	1.46	1.45	1.44
17	1.42	1.51	1.50	1.49	1.47	1.46	1.45	1.44	1.43
18	1.41	1.50	1.49	1.48	1.46	1.45	1.44	1.43	1.42
19	1.41	1.49	1.49	1.47	1.46	1.44	1.43	1.42	1.41
20	1.40	1.49	1.48	1.47	1.45	1.44	1.43	1.42	1.41
21	1.40	1.48	1.48	1.46	1.44	1.43	1.42	1.41	1.40
22	1.40	1.48	1.47	1.45	1.44	1.42	1.41	1.40	1.39
23	1.39	1.47	1.47	1.45	1.43	1.42	1.41	1.40	1.39
24	1.39	1.47	1.46	1.44	1.43	1.41	1.40	1.39	1.38
25	1.39	1.47	1.46	1.44	1.42	1.41	1.40	1.39	1.38
26	1.38	1.46	1.45	1.44	1.42	1.41	1.39	1.38	1.37
27	1.38	1.46	1.45	1.43	1.42	1.40	1.39	1.38	1.37
28	1.38	1.46	1.45	1.43	1.41	1.40	1.39	1.38	1.37
29	1.38	1.45	1.45	1.43	1.41	1.40	1.38	1.37	1.36
30	1.38	1.45	1.44	1.42	1.41	1.39	1.38	1.37	1.36
40	1.36	1.44	1.42	1.40	1.39	1.37	1.36	1.35	1.34
60	1.35	1.42	1.41	1.38	1.37	1.35	1.33	1.32	1.31
120	1.34	1.40	1.39	1.37	1.35	1.33	1.31	1.30	1.29
∞	1.32	1.39	1.37	1.35	1.33	1.31	1.29	1.28	1.27

Table A.5.3.   Upper Percentage Points of F Distribution (α = 0.25) (<u>cont.</u>)

ν_2 \ ν_1:	12	15	20	24	30	40	60	120	∞
1	9.41	9.49	9.58	9.63	9.67	9.71	9.76	9.80	9.85
2	3.39	3.41	3.43	3.43	3.44	3.45	3.46	3.47	3.48
3	2.45	2.46	2.46	2.46	2.47	2.47	2.47	2.47	2.47
4	2.08	2.08	2.08	2.08	2.08	2.08	2.08	2.08	2.08
5	1.89	1.89	1.88	1.88	1.88	1.88	1.87	1.87	1.87
6	1.77	1.76	1.76	1.75	1.75	1.75	1.74	1.74	1.74
7	1.68	1.68	1.67	1.67	1.66	1.66	1.65	1.65	1.65
8	1.62	1.62	1.61	1.60	1.60	1.59	1.59	1.58	1.58
9	1.58	1.57	1.56	1.56	1.55	1.54	1.54	1.53	1.53
10	1.54	1.53	1.52	1.52	1.51	1.51	1.50	1.49	1.48
11	1.51	1.50	1.49	1.49	1.48	1.47	1.47	1.46	1.45
12	1.49	1.48	1.47	1.46	1.45	1.45	1.44	1.43	1.42
13	1.47	1.46	1.45	1.44	1.43	1.42	1.42	1.41	1.40
14	1.45	1.44	1.43	1.42	1.41	1.41	1.40	1.39	1.38
15	1.44	1.43	1.41	1.41	1.40	1.39	1.38	1.37	1.36
16	1.43	1.41	1.40	1.39	1.38	1.37	1.36	1.35	1.34
17	1.41	1.40	1.39	1.38	1.37	1.36	1.35	1.34	1.33
18	1.40	1.39	1.38	1.37	1.36	1.35	1.34	1.33	1.32
19	1.40	1.38	1.37	1.36	1.35	1.34	1.33	1.32	1.30
20	1.39	1.37	1.36	1.35	1.34	1.33	1.32	1.31	1.29
21	1.38	1.37	1.35	1.34	1.33	1.32	1.31	1.30	1.28
22	1.37	1.36	1.34	1.33	1.32	1.31	1.30	1.29	1.28
23	1.37	1.35	1.34	1.33	1.32	1.31	1.30	1.28	1.27
24	1.36	1.35	1.33	1.32	1.31	1.30	1.29	1.28	1.26
25	1.36	1.34	1.33	1.32	1.31	1.29	1.28	1.27	1.25
26	1.35	1.34	1.32	1.31	1.30	1.29	1.28	1.26	1.25
27	1.35	1.33	1.32	1.31	1.30	1.28	1.27	1.26	1.24
28	1.34	1.33	1.31	1.30	1.29	1.28	1.27	1.25	1.24
29	1.34	1.32	1.31	1.30	1.29	1.27	1.26	1.25	1.23
30	1.34	1.32	1.30	1.29	1.28	1.27	1.26	1.24	1.23
40	1.31	1.30	1.28	1.26	1.25	1.24	1.22	1.21	1.19
60	1.29	1.27	1.25	1.24	1.22	1.21	1.19	1.17	1.15
120	1.26	1.24	1.22	1.21	1.19	1.18	1.16	1.13	1.10
∞	1.24	1.22	1.19	1.18	1.16	1.14	1.12	1.08	1.00

Source: Extracted by permission from Pearson and Hartley with corrections by Amos and Pearson (see notes preceding Table A.5.1).

Table A.5.4. Upper Percentage Points of F Distribution (α = 0.10)

ν_2 ν_1:	1	2	3	4	5	6	7	8	9
1	39.9	49.5	53.6	55.8	57.2	58.2	58.9	59.4	59.9
2	8.53	9.00	9.16	9.24	9.29	9.33	9.35	9.37	9.38
3	5.54	5.46	5.39	5.34	5.31	5.28	5.27	5.25	5.24
4	4.54	4.32	4.19	4.11	4.05	4.01	3.98	3.95	3.94
5	4.06	3.78	3.62	3.52	3.45	3.40	3.37	3.34	3.32
6	3.78	3.46	3.29	3.18	3.11	3.05	3.01	2.98	2.96
7	3.59	3.26	3.07	2.96	2.88	2.83	2.78	2.75	2.72
8	3.46	3.11	2.92	2.81	2.73	2.67	2.62	2.59	2.56
9	3.36	3.01	2.81	2.69	2.61	2.55	2.51	2.47	2.44
10	3.29	2.92	2.73	2.61	2.52	2.46	2.41	2.38	2.35
11	3.23	2.86	2.66	2.54	2.45	2.39	2.34	2.30	2.27
12	3.18	2.81	2.61	2.48	2.39	2.33	2.28	2.24	2.21
13	3.14	2.76	2.56	2.43	2.35	2.28	2.23	2.20	2.16
14	3.10	2.73	2.52	2.39	2.31	2.24	2.19	2.15	2.12
15	3.07	2.70	2.49	2.36	2.27	2.21	2.16	2.12	2.09
16	3.05	2.67	2.46	2.33	2.24	2.18	2.13	2.09	2.06
17	3.03	2.64	2.44	2.31	2.22	2.15	2.10	2.06	2.03
18	3.01	2.62	2.42	2.29	2.20	2.13	2.08	2.04	2.00
19	2.99	2.61	2.40	2.27	2.18	2.11	2.06	2.02	1.98
20	2.97	2.59	2.38	2.25	2.16	2.09	2.04	2.00	1.96
21	2.96	2.57	2.36	2.23	2.14	2.08	2.02	1.98	1.95
22	2.95	2.56	2.35	2.22	2.13	2.06	2.01	1.97	1.93
23	2.94	2.55	2.34	2.21	2.11	2.05	1.99	1.95	1.92
24	2.93	2.54	2.33	2.19	2.10	2.04	1.98	1.94	1.91
25	2.92	2.53	2.32	2.18	2.09	2.02	1.97	1.93	1.89
26	2.91	2.52	2.31	2.17	2.08	2.01	1.96	1.92	1.88
27	2.90	2.51	2.30	2.17	2.07	2.00	1.95	1.91	1.87
28	2.89	2.50	2.29	2.16	2.06	2.00	1.94	1.90	1.87
29	2.89	2.50	2.28	2.15	2.06	1.99	1.93	1.89	1.86
30	2.88	2.49	2.28	2.14	2.05	1.98	1.93	1.88	1.85
40	2.84	2.44	2.23	2.09	2.00	1.93	1.87	1.83	1.79
60	2.79	2.39	2.18	2.04	1.95	1.87	1.82	1.77	1.74
120	2.75	2.35	2.13	1.99	1.90	1.82	1.77	1.72	1.68
∞	2.71	2.30	2.08	1.94	1.85	1.77	1.72	1.67	1.63

Table A.5.4.   Upper Percentage Points of F Distribution (α = 0.10) (<u>cont</u>.)

ν_2 ν_1:	12	15	20	24	30	40	60	120	∞
1	60.7	61.2	61.7	62.0	62.3	62.5	62.8	63.1	63.3
2	9.41	9.42	9.44	9.45	9.46	9.47	9.47	9.48	9.49
3	5.22	5.20	5.18	5.18	5.17	5.16	5.15	5.14	5.13
4	3.90	3.87	3.84	3.83	3.82	3.80	3.79	3.78	3.76
5	3.27	3.24	3.21	3.19	3.17	3.16	3.14	3.12	3.10
6	2.90	2.87	2.84	2.82	2.80	2.78	2.76	2.74	2.72
7	2.67	2.63	2.59	2.58	2.56	2.54	2.51	2.49	2.47
8	2.50	2.46	2.42	2.40	2.38	2.36	2.34	2.32	2.29
9	2.38	2.34	2.30	2.28	2.25	2.23	2.21	2.18	2.16
10	2.28	2.24	2.20	2.18	2.16	2.13	2.11	2.08	2.06
11	2.21	2.17	2.12	2.10	2.08	2.05	2.03	2.00	1.97
12	2.15	2.10	2.06	2.04	2.01	1.99	1.96	1.93	1.90
13	2.10	2.05	2.01	1.98	1.96	1.93	1.90	1.88	1.85
14	2.05	2.01	1.96	1.94	1.91	1.89	1.86	1.83	1.80
15	2.02	1.97	1.92	1.90	1.87	1.85	1.82	1.79	1.76
16	1.99	1.94	1.89	1.87	1.84	1.81	1.78	1.75	1.72
17	1.96	1.91	1.86	1.84	1.81	1.78	1.75	1.72	1.69
18	1.93	1.89	1.84	1.81	1.78	1.75	1.72	1.69	1.66
19	1.91	1.86	1.81	1.79	1.76	1.73	1.70	1.67	1.63
20	1.89	1.84	1.79	1.77	1.74	1.71	1.68	1.64	1.61
21	1.87	1.83	1.78	1.75	1.72	1.69	1.66	1.62	1.59
22	1.86	1.81	1.76	1.73	1.70	1.67	1.64	1.60	1.57
23	1.84	1.80	1.74	1.72	1.69	1.66	1.62	1.59	1.55
24	1.83	1.78	1.73	1.70	1.67	1.64	1.61	1.57	1.53
25	1.82	1.77	1.72	1.69	1.66	1.63	1.59	1.56	1.52
26	1.81	1.76	1.71	1.68	1.65	1.61	1.58	1.54	1.50
27	1.80	1.75	1.70	1.67	1.64	1.60	1.57	1.53	1.49
28	1.79	1.74	1.69	1.66	1.63	1.59	1.56	1.52	1.48
29	1.78	1.73	1.68	1.65	1.62	1.58	1.55	1.51	1.47
30	1.77	1.72	1.67	1.64	1.61	1.57	1.54	1.50	1.46
40	1.71	1.66	1.61	1.57	1.54	1.51	1.47	1.42	1.38
60	1.66	1.60	1.54	1.51	1.48	1.44	1.40	1.35	1.29
120	1.60	1.55	1.48	1.45	1.41	1.37	1.32	1.26	1.19
∞	1.55	1.49	1.42	1.38	1.34	1.30	1.24	1.17	1.00

     Source: Extracted by permission from Pearson and Hartley with correc-
tions by Amos and Pearson (see notes preceding Table A.5.1).

Table A.5.5. Upper Percentage Points of F Distribution (α = 0.025)

ν_2 \ ν_1:	1	2	3	4	5	6	7	8	9
1	648	800	864	900	922	937	948	957	963
2	38.5	39.0	39.2	39.2	39.3	39.3	39.4	39.4	39.4
3	17.4	16.0	15.4	15.1	14.9	14.7	14.6	14.5	14.5
4	12.2	10.6	9.98	9.60	9.36	9.20	9.07	8.98	8.90
5	10.0	8.43	7.76	7.39	7.15	6.98	6.85	6.76	6.68
6	8.81	7.26	6.60	6.23	5.99	5.82	5.70	5.60	5.52
7	8.07	6.54	5.89	5.52	5.29	5.12	4.99	4.90	4.82
8	7.57	6.06	5.42	5.05	4.82	4.65	4.53	4.43	4.36
9	7.21	5.71	5.08	4.72	4.48	4.32	4.20	4.10	4.03
10	6.94	5.46	4.83	4.47	4.24	4.07	3.95	3.85	3.78
11	6.72	5.26	4.63	4.28	4.04	3.88	3.76	3.66	3.59
12	6.55	5.10	4.47	4.12	3.89	3.73	3.61	3.51	3.44
13	6.41	4.97	4.35	4.00	3.77	3.60	3.48	3.39	3.31
14	6.30	4.86	4.24	3.89	3.66	3.50	3.38	3.29	3.21
15	6.20	4.77	4.15	3.80	3.58	3.41	3.29	3.20	3.12
16	6.12	4.69	4.08	3.73	3.50	3.34	3.22	3.12	3.05
17	6.04	4.62	4.01	3.66	3.44	3.28	3.16	3.06	2.98
18	5.98	4.56	3.95	3.61	3.38	3.22	3.10	3.01	2.93
19	5.92	4.51	3.90	3.56	3.33	3.17	3.05	2.96	2.88
20	5.87	4.46	3.86	3.51	3.29	3.13	3.01	2.91	2.84
21	5.83	4.42	3.82	3.48	3.25	3.09	2.97	2.87	2.80
22	5.79	4.38	3.78	3.44	3.22	3.05	2.93	2.84	2.76
23	5.75	4.35	3.75	3.41	3.18	3.02	2.90	2.81	2.73
24	5.72	4.32	3.72	3.38	3.15	2.99	2.87	2.78	2.70
25	5.69	4.29	3.69	3.35	3.13	2.97	2.85	2.75	2.68
26	5.66	4.27	3.67	3.33	3.10	2.94	2.82	2.73	2.65
27	5.63	4.24	3.65	3.31	3.08	2.92	2.80	2.71	2.63
28	5.61	4.22	3.63	3.29	3.06	2.90	2.78	2.69	2.61
29	5.59	4.20	3.61	3.27	3.04	2.88	2.76	2.67	2.59
30	5.57	4.18	3.59	3.25	3.03	2.87	2.75	2.65	2.57
40	5.42	4.05	3.46	3.13	2.90	2.74	2.62	2.53	2.45
60	5.29	3.93	3.34	3.01	2.79	2.63	2.51	2.41	2.33
120	5.15	3.80	3.23	2.89	2.67	2.52	2.39	2.30	2.22
∞	5.02	3.69	3.12	2.79	2.57	2.41	2.29	2.19	2.11

Table A.5.5.   Upper Percentage Points of F Distribution (α = 0.025) (<u>cont</u>.)

ν_2 \ ν_1:	12	15	20	24	30	40	60	120	∞
1	977	985	993	997	1001	1006	1010	1014	1018
2	39.4	39.4	39.4	39.5	39.5	39.5	39.5	39.5	39.5
3	14.3	14.2	14.2	14.1	14.1	14.0	14.0	14.0	13.9
4	8.75	8.66	8.56	8.51	8.46	8.41	8.36	8.31	8.26
5	6.52	6.43	6.33	6.28	6.23	6.18	6.12	6.07	6.02
6	5.37	5.27	5.17	5.12	5.07	5.01	4.96	4.90	4.85
7	4.67	4.57	4.47	4.42	4.36	4.31	4.25	4.20	4.14
8	4.20	4.10	4.00	3.95	3.89	3.84	3.78	3.73	3.67
9	3.87	3.77	3.67	3.61	3.56	3.51	3.45	3.39	3.33
10	3.62	3.52	3.42	3.37	3.31	3.26	3.20	3.14	3.08
11	3.43	3.33	3.23	3.17	3.12	3.06	3.00	2.94	2.88
12	3.28	3.18	3.07	3.02	2.96	2.91	2.85	2.79	2.72
13	3.15	3.05	2.95	2.89	2.84	2.78	2.72	2.66	2.60
14	3.05	2.95	2.84	2.79	2.73	2.67	2.61	2.55	2.49
15	2.96	2.86	2.76	2.70	2.64	2.59	2.52	2.46	2.40
16	2.89	2.79	2.68	2.63	2.57	2.51	2.45	2.38	2.32
17	2.82	2.72	2.62	2.56	2.50	2.44	2.38	2.32	2.25
18	2.77	2.67	2.56	2.50	2.44	2.38	2.32	2.26	2.19
19	2.72	2.62	2.51	2.45	2.39	2.33	2.27	2.20	2.13
20	2.68	2.57	2.46	2.41	2.35	2.29	2.22	2.16	2.09
21	2.64	2.53	2.42	2.37	2.31	2.25	2.18	2.11	2.04
22	2.60	2.50	2.39	2.33	2.27	2.21	2.14	2.08	2.00
23	2.57	2.47	2.36	2.30	2.24	2.18	2.11	2.04	1.97
24	2.54	2.44	2.33	2.27	2.21	2.15	2.08	2.01	1.94
25	2.51	2.41	2.30	2.24	2.18	2.12	2.05	1.98	1.91
26	2.49	2.39	2.28	2.22	2.16	2.09	2.03	1.95	1.88
27	2.47	2.36	2.25	2.19	2.13	2.07	2.00	1.93	1.85
28	2.45	2.34	2.23	2.17	2.11	2.05	1.98	1.91	1.83
29	2.43	2.32	2.21	2.15	2.09	2.03	1.96	1.89	1.81
30	2.41	2.31	2.20	2.14	2.07	2.01	1.94	1.87	1.79
40	2.29	2.18	2.07	2.01	1.94	1.88	1.80	1.72	1.64
60	2.17	2.06	1.94	1.88	1.82	1.74	1.67	1.58	1.48
120	2.05	1.94	1.82	1.76	1.69	1.61	1.53	1.43	1.31
∞	1.94	1.83	1.71	1.64	1.57	1.48	1.39	1.27	1.00

Source: Extracted by permission from Pearson and Hartley with corrections by Amos and Pearson (see notes preceding Table A.5.1).

Table A.5.6. Upper Percentage Points of F Distribution ($\alpha = 0.005$)

ν_2 \ ν_1:	1	2	3	4	5	6	7	8	9
1	162*	200*	216*	225*	231*	234*	237*	239*	241*
2	198	199	199	199	199	199	199	199	199
3	55.5	49.8	47.5	46.2	45.4	44.8	44.4	44.1	43.9
4	31.3	26.3	24.3	23.2	22.5	22.0	21.6	21.4	21.1
5	22.8	18.3	16.5	15.6	14.9	14.5	14.2	14.0	13.8
6	18.6	14.5	12.9	12.0	11.5	11.1	10.8	10.6	10.4
7	16.2	12.4	10.9	10.0	9.52	9.16	8.89	8.68	8.51
8	14.7	11.0	9.60	8.81	8.30	7.95	7.69	7.50	7.34
9	13.6	10.1	8.72	7.96	7.47	7.13	6.88	6.69	6.54
10	12.8	9.43	8.08	7.34	6.87	6.54	6.30	6.12	5.97
11	12.2	8.91	7.60	6.88	6.42	6.10	5.86	5.68	5.54
12	11.8	8.51	7.23	6.52	6.07	5.76	5.52	5.35	5.20
13	11.4	8.19	6.93	6.23	5.79	5.48	5.25	5.08	4.94
14	11.1	7.92	6.68	6.00	5.56	5.26	5.03	4.86	4.72
15	10.8	7.70	6.48	5.80	5.37	5.07	4.85	4.67	4.54
16	10.6	7.51	6.30	5.64	5.21	4.91	4.69	4.52	4.38
17	10.4	7.35	6.16	5.50	5.07	4.78	4.56	4.39	4.25
18	10.2	7.21	6.03	5.37	4.96	4.66	4.44	4.28	4.14
19	10.1	7.09	5.92	5.27	4.85	4.56	4.34	4.18	4.04
20	9.94	6.99	5.82	5.17	4.76	4.47	4.26	4.09	3.96
21	9.83	6.89	5.73	5.09	4.68	4.39	4.18	4.01	3.88
22	9.73	6.81	5.65	5.02	4.61	4.32	4.11	3.94	3.81
23	9.63	6.73	5.58	4.95	4.54	4.26	4.05	3.88	3.75
24	9.55	6.66	5.52	4.89	4.49	4.20	3.99	3.83	3.69
25	9.48	6.60	5.46	4.84	4.43	4.15	3.94	3.78	3.64
26	9.41	6.54	5.41	4.79	4.38	4.10	3.89	3.73	3.60
27	9.34	6.49	5.36	4.74	4.34	4.06	3.85	3.69	3.56
28	9.28	6.44	5.32	4.70	4.30	4.02	3.81	3.65	3.52
29	9.23	6.40	5.28	4.66	4.26	3.98	3.77	3.61	3.48
30	9.18	6.35	5.24	4.62	4.23	3.95	3.74	3.58	3.45
40	8.83	6.07	4.98	4.37	3.99	3.71	3.51	3.35	3.22
60	8.49	5.79	4.73	4.14	3.76	3.49	3.29	3.13	3.01
120	8.18	5.54	4.50	3.92	3.55	3.28	3.09	2.93	2.81
∞	7.88	5.30	4.28	3.72	3.35	3.09	2.90	2.74	2.62

*Multiply these entries by 100.

Table A.5.6. Upper Percentage Points of F Distribution (α = 0.005) (<u>cont.</u>)

ν_2 ν_1:	12	15	20	24	30	40	60	120	∞
1	244*	246*	248*	249*	250*	251*	253*	254*	255*
2	199	199	199	200	200	200	200	200	200
3	43.4	43.1	42.8	42.6	42.5	42.3	42.2	42.0	41.8
4	20.7	20.4	20.2	20.0	19.9	19.8	19.6	19.5	19.3
5	13.4	13.2	12.9	12.8	12.7	12.5	12.4	12.3	12.1
6	10.0	9.81	9.59	9.47	9.36	9.24	9.12	9.00	8.88
7	8.18	7.97	7.75	7.65	7.53	7.42	7.31	7.19	7.08
8	7.01	6.81	6.61	6.50	6.40	6.29	6.18	6.06	5.95
9	6.23	6.03	5.83	5.73	5.62	5.52	5.41	5.30	5.19
10	5.66	5.47	5.27	5.17	5.07	4.97	4.86	4.75	4.64
11	5.24	5.05	4.86	4.76	4.65	4.55	4.44	4.34	4.23
12	4.91	4.72	4.53	4.43	4.33	4.23	4.12	4.01	3.90
13	4.64	4.46	4.27	4.17	4.07	3.97	3.87	3.76	3.65
14	4.43	4.25	4.06	3.96	3.86	3.76	3.66	3.55	3.44
15	4.25	4.07	3.88	3.79	3.69	3.58	3.48	3.37	3.26
16	4.10	3.92	3.73	3.64	3.54	3.44	3.33	3.22	3.11
17	3.97	3.79	3.61	3.51	3.41	3.31	3.21	3.10	2.98
18	3.86	3.68	3.50	3.40	3.30	3.20	3.10	2.99	2.87
19	3.76	3.59	3.40	3.31	3.21	3.11	3.00	2.89	2.78
20	3.68	3.50	3.32	3.22	3.12	3.02	2.92	2.81	2.69
21	3.60	3.43	3.24	3.15	3.05	2.95	2.84	2.73	2.61
22	3.54	3.36	3.18	3.08	2.98	2.88	2.77	2.66	2.55
23	3.47	3.30	3.12	3.02	2.92	2.82	2.71	2.60	2.48
24	3.42	3.25	3.06	2.97	2.87	2.77	2.66	2.55	2.43
25	3.37	3.20	3.01	2.92	2.82	2.72	2.61	2.50	2.38
26	3.33	3.15	2.97	2.87	2.77	2.67	2.56	2.45	2.33
27	3.28	3.11	2.93	2.83	2.73	2.63	2.52	2.41	2.29
28	3.25	3.07	2.89	2.79	2.69	2.59	2.48	2.37	2.25
29	3.21	3.04	2.86	2.76	2.66	2.56	2.45	2.33	2.21
30	3.18	3.01	2.82	2.73	2.63	2.52	2.42	2.30	2.18
40	2.95	2.78	2.60	2.50	2.40	2.30	2.18	2.06	1.93
60	2.74	2.57	2.39	2.29	2.19	2.08	1.96	1.83	1.69
120	2.54	2.37	2.19	2.09	1.98	1.87	1.75	1.61	1.43
∞	2.36	2.19	2.00	1.90	1.79	1.67	1.53	1.36	1.00

Source: Extracted by permission from Pearson and Hartley with corrections by Amos and Pearson (see notes preceding Table A.5.1).
*Multiply these entries by 100.

Table A.5.7. Upper Percentage Points of F Distribution (α = 0.001)

ν_2 \ ν_1:	1	2	3	4	5	6	7	8	9
1	405*	500*	540*	562*	576*	586*	593*	598*	602*
2	998	999	999	999	999	999	999	999	999
3	167	148	141	137	135	133	132	131	130
4	74.1	61.2	56.2	53.4	51.7	50.5	49.7	49.0	48.5
5	47.2	37.1	33.2	31.1	29.8	28.8	28.2	27.6	27.2
6	35.5	27.0	23.7	21.9	20.8	20.0	19.5	19.0	18.7
7	29.2	21.7	18.8	17.2	16.2	15.5	15.0	14.6	14.3
8	25.4	18.5	15.8	14.4	13.5	12.9	12.4	12.0	11.8
9	22.9	16.4	13.9	12.6	11.7	11.1	10.7	10.4	10.1
10	21.0	14.9	12.6	11.3	10.5	9.92	9.52	9.20	8.96
11	19.7	13.8	11.6	10.4	9.58	9.05	8.66	8.35	8.12
12	18.6	13.0	10.8	9.63	8.89	8.38	8.00	7.71	7.48
13	17.8	12.3	10.2	9.07	8.35	7.86	7.49	7.21	6.98
14	17.1	11.8	9.73	8.62	7.92	7.43	7.08	6.80	6.58
15	16.6	11.3	9.34	8.25	7.57	7.09	6.74	6.47	6.26
16	16.1	11.0	9.00	7.94	7.27	6.81	6.46	6.19	5.98
17	15.7	10.7	8.73	7.68	7.02	6.56	6.22	5.96	5.75
18	15.4	10.4	8.49	7.46	6.81	6.35	6.02	5.76	5.56
19	15.1	10.2	8.28	7.26	6.62	6.18	5.85	5.59	5.39
20	14.8	9.95	8.10	7.10	6.46	6.02	5.69	5.44	5.24
21	14.6	9.77	7.94	6.95	6.32	5.88	5.56	5.31	5.11
22	14.4	9.61	7.80	6.81	6.19	5.76	5.44	5.19	4.99
23	14.2	9.47	7.67	6.69	6.08	5.65	5.33	5.09	4.89
24	14.0	9.34	7.55	6.59	5.98	5.55	5.23	4.99	4.80
25	13.9	9.22	7.45	6.49	5.88	5.46	5.15	4.91	4.71
26	13.7	9.12	7.36	6.41	5.80	5.38	5.07	4.83	4.64
27	13.6	9.02	7.27	6.33	5.73	5.31	5.00	4.76	4.57
28	13.5	8.93	7.19	6.25	5.66	5.24	4.93	4.69	4.50
29	13.4	8.85	7.12	6.19	5.59	5.18	4.87	4.64	4.45
30	13.3	8.77	7.05	6.12	5.53	5.12	4.82	4.58	4.39
40	12.6	8.25	6.60	5.70	5.13	4.73	4.44	4.21	4.02
60	12.0	7.76	6.17	5.31	4.76	4.37	4.09	3.87	3.69
120	11.4	7.32	5.79	4.95	4.42	4.04	3.77	3.55	3.38
∞	10.8	6.91	5.42	4.62	4.10	3.74	3.47	3.27	3.10

*Multiply these entries by 1000.

Table A.5.7.   Upper Percentage Points of F Distribution (α = 0.001) (<u>cont.</u>)

ν_2 ν_1:	12	15	20	24	30	40	60	120	∞
1	611*	616*	621*	624*	626*	629*	631*	634*	637*
2	999	999	999	999	999	999	999	999	999
3	128	127	126	126	125	125	124	124	124
4	47.4	46.8	46.1	45.8	45.4	45.1	44.8	44.4	44.0
5	26.4	25.9	25.4	25.1	24.9	24.6	24.3	24.1	23.8
6	18.0	17.6	17.1	16.9	16.7	16.4	16.2	16.0	15.8
7	13.7	13.3	12.9	12.7	12.5	12.3	12.1	11.9	11.7
8	11.2	10.8	10.5	10.3	10.1	9.92	9.73	9.53	9.33
9	9.57	9.24	8.90	8.72	8.55	8.37	8.19	8.00	7.81
10	8.45	8.13	7.80	7.64	7.47	7.30	7.12	6.94	6.76
11	7.63	7.32	7.01	6.85	6.68	6.52	6.35	6.17	6.00
12	7.00	6.71	6.40	6.25	6.09	5.93	5.76	5.59	5.42
13	6.52	6.23	5.93	5.78	5.63	5.47	5.30	5.14	4.97
14	6.13	5.85	5.56	5.41	5.25	5.10	4.94	4.77	4.60
15	5.81	5.54	5.25	5.10	4.95	4.80	4.64	4.47	4.31
16	5.55	5.27	4.99	4.85	4.70	4.54	4.39	4.23	4.06
17	5.32	5.05	4.78	4.63	4.48	4.33	4.18	4.02	3.85
18	5.13	4.87	4.59	4.45	4.30	4.15	4.00	3.84	3.67
19	4.97	4.70	4.43	4.29	4.14	3.99	3.84	3.68	3.51
20	4.82	4.56	4.29	4.15	4.00	3.86	3.70	3.54	3.38
21	4.70	4.44	4.17	4.03	3.88	3.74	3.58	3.42	3.26
22	4.58	4.33	4.06	3.92	3.78	3.63	3.48	3.32	3.15
23	4.48	4.23	3.96	3.82	3.68	3.53	3.38	3.22	3.05
24	4.39	4.14	3.87	3.74	3.59	3.45	3.29	3.14	2.97
25	4.31	4.06	3.79	3.66	3.52	3.37	3.22	3.06	2.89
26	4.24	3.99	3.72	3.59	3.44	3.30	3.15	2.99	2.82
27	4.17	3.92	3.66	3.52	3.38	3.23	3.08	2.92	2.75
28	4.11	3.86	3.60	3.46	3.32	3.18	3.02	2.86	2.69
29	4.05	3.80	3.54	3.41	3.27	3.12	2.97	2.81	2.64
30	4.00	3.75	3.49	3.36	3.22	3.07	2.92	2.76	2.59
40	3.64	3.40	3.15	3.01	2.87	2.73	2.57	2.41	2.23
60	3.31	3.08	2.83	2.69	2.55	2.41	2.25	2.08	1.89
120	3.02	2.78	2.53	2.40	2.26	2.11	1.95	1.76	1.54
∞	2.74	2.51	2.27	2.13	1.99	1.84	1.66	1.45	1.00

Source: Extracted by permission from Pearson and Hartley with corrections by Amos and Pearson (see notes preceding Table A.5.1).
   *Multiply these entries by 1000.

TABLE A.6.   UPPER PERCENTAGE POINTS OF F_{max} DISTRIBUTION (1–CDF)

Table A.6.1.   Upper Percentage Points of F_{max} Distribution (1–CDF):   Equal Replication

ν	α	t=3	4	5	6	7	8	9	10	12
2	.25	15.4	25.1	35.5	46.8	58.5	70.8	83.6	96.7	124
	.10	42.5	69.1	98.2	129	162	196	231	267	342
	.05	87.5	142	202	266	333	403	475	550	704
	.01	448	729	1036	1362	1705	2063	2432	2813	3605
3	.25	8.13	11.7	15.1	18.5	21.9	25.2	28.4	31.6	37.8
	.10	16.8	24.0	30.9	37.7	44.4	50.9	57.4	63.7	76.1
	.05	27.8	39.2	50.7	62.0	72.9	83.5	93.9	104	124
	.01	85	120	151	184	216	249	281	310	361
4	.25	5.79	7.81	9.67	11.4	13.1	14.7	16.2	17.6	20.4
	.10	10.4	13.9	17.1	20.1	22.9	25.6	28.1	30.6	35.3
	.05	15.5	20.6	25.2	29.5	33.6	37.5	41.1	44.6	51.4
	.01	37	49	59	69	79	89	97	106	120
5	.25	4.66	6.05	7.28	8.41	9.46	10.4	11.4	12.2	13.9
	.10	7.68	9.86	11.8	13.5	15.2	16.7	18.1	19.4	21.9
	.05	10.8	13.7	16.3	18.7	20.8	22.9	24.7	26.5	29.9
	.01	22	28	33	38	42	46	50	54	60
6	.25	4.00	5.05	5.97	6.78	7.53	8.22	8.87	9.48	10.6
	.10	6.23	7.78	9.11	10.3	11.4	12.4	13.3	14.2	15.8
	.05	8.38	10.4	12.1	13.7	15.0	16.3	17.5	18.6	20.7
	.01	15.5	19.1	22	25	27	30	32	34	37
7	.25	3.57	4.41	5.14	5.77	6.35	6.88	7.37	7.82	8.66
	.10	5.32	6.52	7.52	8.41	9.20	9.93	10.6	11.2	12.4
	.05	6.94	8.44	9.70	10.8	11.8	12.7	13.5	14.3	15.8
	.01	12.1	14.5	16.5	18.4	20	22	23	24	27
8	.25	3.26	3.97	4.57	5.09	5.55	5.98	6.37	6.73	7.40
	.10	4.71	5.68	6.48	7.18	7.80	8.36	8.88	9.36	10.2
	.05	6.00	7.18	8.12	9.03	9.78	10.5	11.1	11.7	12.7
	.01	9.9	11.7	13.2	14.5	15.8	16.9	17.9	18.9	21
9	.25	3.03	3.64	4.15	4.59	4.98	5.34	5.66	5.96	6.51
	.10	4.26	5.07	5.74	6.32	6.82	7.28	7.70	8.09	8.78
	.05	5.34	6.31	7.11	7.80	8.41	8.95	9.45	9.91	10.7
	.01	8.5	9.9	11.1	12.1	13.1	13.9	14.7	15.3	16.6

$f_{max, \alpha, t, \nu}$

use $\alpha = .25$

Table A.6.1. Upper Percentage Points of F_{max} Distribution (1-CDF): Equal
              Replication (<u>cont.</u>)

ν	α	t=3	4	5	6	7	8	9	10	12
10	.25	2.85	3.39	3.83	4.21	4.55	4.86	5.13	5.39	5.85
	.10	3.93	4.63	5.19	5.68	6.11	6.49	6.84	7.16	7.74
	.05	4.85	5.67	6.34	6.92	7.42	7.87	8.28	8.66	9.34
	.01	7.4	8.6	9.6	10.4	11.1	11.8	12.4	12.9	13.9
12	.25	2.58	3.02	3.38	3.68	3.95	4.18	4.40	4.60	4.95
	.10	3.45	4.00	4.44	4.81	5.13	5.42	5.68	5.92	6.35
	.05	4.16	4.79	5.30	5.72	6.09	6.42	6.72	7.00	7.48
	.01	6.1	6.9	7.6	8.2	8.7	9.1	9.5	9.9	10.6
15	.25	2.32	2.67	2.95	3.18	3.38	3.56	3.72	3.87	4.13
	.10	3.00	3.41	3.74	4.02	4.25	4.46	4.65	4.82	5.13
	.05	3.54	4.01	4.37	4.68	4.95	5.19	5.40	5.59	5.93
	.01	4.9	5.5	6.0	6.4	6.7	7.1	7.3	7.5	8.0
20	.25	2.07	2.33	2.53	2.70	2.85	2.98	3.09	3.20	3.38
	.10	2.57	2.87	3.10	3.29	3.46	3.60	3.73	3.85	4.06
	.05	2.95	3.29	3.54	3.76	3.94	4.10	4.24	4.37	4.59
	.01	3.8	4.3	4.6	4.9	5.1	5.3	5.5	5.6	5.9
30	.25	1.80	1.98	2.12	2.24	2.34	2.42	2.49	2.56	2.68
	.10	2.14	2.34	2.50	2.62	2.73	2.82	2.90	2.97	3.10
	.05	2.40	2.61	2.78	2.91	3.02	3.12	3.21	3.29	3.39
	.01	3.0	3.3	3.4	3.6	3.7	3.8	3.9	4.0	4.2
60	.25	1.51	1.62	1.70	1.76	1.81	1.86	1.90	1.93	2.00
	.10	1.71	1.82	1.90	1.96	2.02	2.07	2.11	2.14	2.21
	.05	1.85	1.96	2.04	2.11	2.17	2.22	2.26	2.30	2.36
	.01	2.2	2.3	2.4	2.4	2.5	2.5	2.6	2.6	2.7
∞		1.00	1.00	1.00	1.00	1.00	1.00	1.00	1.00	1.00

Source: Values for α = 0.25 and 0.10 extracted by permission from R. J.
Beckman and G. L. Tietjen, *Biometrika* 60(1973):213-14. Values for α = 0.05
and 0.01 extracted by permission from H. A. David, *Biometrika* 39(1952):422-24.
Values of $f_{max,\alpha,t,\nu}$ are used as critical values in testing the homogeneity of
variances from t independent groups with ν df per group. The test statistic
is $f_{max} = s^2_{max}/s^2_{min}$.

Table A.6.2. Upper 5% Points of F_{max} Distribution (1-CDF): Unequal Replication

ν_2 for t_2 of t groups	t_1, t_2	ν_1 for t_1 of t groups $(t_1+t_2=t)$							
		2	4	6	8	10	12	14	16
2	1,2	87.5	58.4	55.7	55.0	54.7	54.5	54.4	54.2
	1,3	142	105	101	100	99.9	99.5	99.3	99.1
	2,2	142	70.1	63.7	61.9	60.8	60.1	59.6	59.1
	1,4	202	158	154	153	152	152	152	151
	2,3	202	116	110	107	106	105	104	104
	1,5	266	216	213	211	210	210	209	209
	2,4	266	169	162	160	158	157	156	155
	3,3	266	126	116	112	110	109	108	107
4	1,2	33.0	15.5	13.0	12.2	11.8	11.6	11.5	11.4
	1,3	40.9	20.6	17.9	17.1	16.8	16.7	16.6	16.5
	2,2	70.1	20.6	15.3	13.7	13.0	12.6	12.4	12.3
	1,4	47.4	25.2	22.5	21.8	21.5	21.4	21.3	21.3
	2,3	78.9	25.2	19.8	18.3	17.7	17.4	17.2	17.1
	1,5	53.1	29.5	26.9	26.2	25.9	25.8	25.8	25.7
	2,4	86.3	29.5	24.1	22.8	22.2	22.0	21.8	21.7
	3,3	126	29.5	21.4	19.2	18.4	18.0	17.7	17.6
6	1,2	26.9	10.6	8.38	7.57	7.18	6.97	6.84	6.75
	1,3	31.0	12.8	10.4	9.58	9.23	9.04	8.93	8.87
	2,2	63.7	15.3	10.4	8.78	8.04	7.64	7.39	7.24
	1,4	34.1	14.6	12.1	11.3	11.0	10.8	10.8	10.7
	2,3	69.1	17.2	12.1	10.5	9.86	9.52	9.32	9.20
	1,5	36.6	16.1	13.7	12.9	12.6	12.5	12.4	12.3
	2,4	73.2	18.7	13.7	12.1	11.5	11.2	11.0	10.9
	3,3	116	21.4	13.7	11.4	10.4	9.90	9.62	9.45
8	1,2	25.1	8.91	6.78	6.00	5.62	5.39	5.25	5.15
	1,3	28.3	10.2	7.98	7.18	6.81	6.61	6.48	6.40
	2,2	61.9	13.7	8.78	7.18	6.43	6.01	5.75	5.57
	1,4	30.6	11.3	8.95	8.12	7.81	7.62	7.52	7.45
	2,3	66.3	14.8	9.75	8.12	7.45	7.06	6.83	6.69
	1,5	32.4	12.2	9.79	9.03	8.68	8.51	8.41	8.36
	2,4	69.5	15.7	10.6	9.03	8.33	7.98	7:78	7.66
	3,3	112	19.2	11.4	9.03	7.98	7.44	7.11	6.91

Table A.6.2. Upper 5% Points of F_{max} Distribution (1-CDF): Unequal
Replication (<u>cont.</u>)

ν_2 for t_2 of t groups	t_1, t_2	ν_1 for t_1 of t groups $(t_1+t_2)=t$							
		2	4	6	8	10	12	14	16
10	1,2	24.4	8.11	5.99	5.23	4.85	4.62	4.47	4.37
	1,3	27.1	9.09	6.84	6.05	5.67	5.45	5.32	5.23
	2,2	60.8	13.0	8.04	6.43	5.67	5.23	4.96	4.77
	1,4	29.1	9.83	7.50	6.71	6.34	6.14	6.01	5.94
	2,3	64.7	13.8	8.69	7.08	6.34	5.93	5.68	5.52
	1,5	30.6	10.4	8.06	7.27	6.92	6.72	6.61	6.54
	2,4	67.5	14.4	9.23	7.63	6.92	6.52	6.30	6.15
	3,3	110	18.4	10.4	7.98	6.92	6.33	5.98	5.75
12	1,2	23.9	7.66	5.54	4.78	4.39	4.16	4.01	3.90
	1,3	26.4	8.45	6.19	5.40	5.01	4.79	4.65	4.55
	2,2	60.1	12.6	7.64	6.01	5.23	4.79	4.51	4.31
	1,4	28.1	9.04	6.69	5.89	5.51	5.30	5.16	5.08
	2,3	63.6	13.3	8.13	6.48	5.72	5.30	5.03	4.86
	1,5	29.5	9.50	7.10	6.30	5.93	5.72	5.60	5.52
	2,4	66.1	13.8	8.52	6.87	6.13	5.72	5.47	5.31
	3,3	109	18.0	9.90	7.44	6.33	5.72	5.35	5.09
14	1,2	23.6	7.37	5.24	4.48	4.09	3.86	3.73	3.60
	1,3	25.9	8.06	5.78	4.98	4.59	4.36	4.24	4.12
	2,2	59.6	12.4	7.39	5.75	4.96	4.51	4.24	4.02
	1,4	27.4	8.56	6.18	5.37	4.98	4.76	4.64	4.53
	2,3	62.7	13.0	7.79	6.12	5.33	4.90	4.64	4.44
	1,5	28.7	8.94	6.51	5.69	5.31	5.09	4.98	4.87
	2,4	65.0	13.4	8.10	6.42	5.64	5.22	4.98	4.79
	3,3	108	17.7	9.62	7.11	5.98	5.35	4.98	4.73
16	1,2	23.3	7.18	5.04	4.27	3.88	3.65	3.49	3.40
	1,3	25.4	7.80	5.50	4.70	4.30	4.07	3.92	3.84
	2,2	59.1	12.3	7.24	5.57	4.77	4.31	4.02	3.84
	1,4	26.9	8.24	5.84	5.02	4.62	4.39	4.25	4.17
	2,3	62.1	12.8	7.57	5.87	5.08	4.63	4.34	4.17
	1,5	28.0	8.58	6.11	5.28	4.88	4.66	4.52	4.46
	2,4	64.2	13.2	7.84	6.12	5.32	4.88	4.61	4.46
	3,3	107	17.6	9.45	6.91	5.75	5.09	4.73	4.46

Source: Values extracted by permission from N. A. Hartman, Jr., 1969
(Tech. Rep. 15, Dept. Stat.) Oregon State Univ., Corvallis. These values are
useful in testing homogeneity of variances among t = 6 or fewer groups con-
sisting of t_1 groups with ν_1 df each and t_2 groups with ν_2 df each ($t_1+t_2 = t$;
ν_1 and $\nu_2 \le 16$ each).

TABLE A.7.   COEFFICIENTS OF ORTHOGONAL POLYNOMIALS

Table A.7.1.   Coefficients of Orthogonal Polynomials (ξ'_{ij}):   Equally Spaced Levels (i = treatment level, j = polynomial degree)

i	t=3 j=1	2	t=4 j=1	2	3	t=5 j=1	2	3	4
1	-1	+1	-3	+1	-1	-2	+2	-1	+1
2	0	-2	-1	-1	+3	-1	-1	+2	-4
3	+1	+1	+1	-1	-3	0	-2	0	+6
4			+3	+1	+1	+1	-1	-2	-4
5						+2	+2	+1	+1
$\sum_{i=1}^{t}(\xi'_{ij})^2$	2	6	20	4	20	10	14	10	70
m_j	1	3	2	1	10/3	1	1	5/6	35/12

i	t=6 j=1	2	3	4	5	t=7 j=1	2	3	4	5
1	-5	+5	-5	+1	-1	-3	+5	-1	+3	-1
2	-3	-1	+7	-3	+5	-2	0	+1	-7	+4
3	-1	-4	+4	+2	-10	-1	-3	+1	+1	-5
4	+1	-4	-4	+2	+10	0	-4	0	+6	0
5	+3	-1	-7	-3	-5	+1	-3	-1	+1	+5
6	+5	+5	+5	+1	+1	+2	0	-1	-7	-4
7						+3	+5	+1	+3	+1
$\sum_{i=1}^{t}(\xi'_{ij})^2$	70	84	180	28	252	28	84	6	154	84
m_j	2	3/2	5/3	7/12	21/10	1	1	1/6	7/12	7/20

i	t=8 j=1	2	3	4	5	t=9 j=1	2	3	4	5
1	-7	+7	-7	+7	-7	-4	+28	-14	+14	-4
2	-5	+1	+5	-13	+23	-3	+7	+7	-21	+11
3	-3	-3	+7	-3	-17	-2	-8	+13	-11	-4
4	-1	-5	+3	+9	-15	-1	-17	+9	+9	-9
5	+1	-5	-3	+9	+15	0	-20	0	+18	0
6	+3	-3	-7	-3	+17	+1	-17	-9	+9	+9
7	+5	+1	-5	-13	-23	+2	-8	-13	-11	+4
8	+7	+7	+7	+7	+7	+3	+7	-7	-21	-11
9						+4	+28	+14	+14	+4
$\sum_{i=1}^{t}(\xi'_{ij})^2$	168	168	264	616	2184	60	2772	990	2002	468
m_j	2	1	2/3	7/12	7/10	1	3	5/6	7/12	3/20

Table A.7.1. Coefficients of Orthogonal Polynomials (ξ'_{ij}): Equally Spaced Levels (<u>cont.</u>)

i	t=10* j=1	2	3	4	5	t=11* j=1	2	3	4	5
1	-9	+6	-42	+18	-6	-5	+15	-30	+6	-3
2	-7	+2	+14	-22	+14	-4	+6	+6	-6	+6
3	-5	-1	+35	-17	-1	-3	-1	+22	-6	+1
4	-3	-3	+31	+3	-11	-2	-6	+23	-1	-4
5	-1	-4	+12	+18	-6	-1	-9	+14	+4	-4
6						0	-10	0	+6	0
$\sum_{i=1}^{t}(\xi'_{ij})^2$	330	132	8580	2860	780	110	858	4290	286	156
m_j	2	1/2	5/3	5/12	1/10	1	1	5/6	1/12	1/40

i	t=12* j=1	2	3	4	5	t=13* j=1	2	3	4	5
1	-11	+55	-33	+33	-33	-6	+22	-11	+99	-22
2	-9	+25	+3	-27	+57	-5	+11	0	-66	+33
3	-7	+1	+21	-33	+21	-4	+2	+6	-96	+18
4	-5	-17	+25	-13	-29	-3	-5	+8	-54	-11
5	-3	-29	+19	+12	-44	-2	-10	+7	+11	-26
6	-1	-35	+7	+28	-20	-1	-13	+4	+64	-20
7						0	-14	0	+84	0
$\sum_{i=1}^{t}(\xi'_{ij})^2$	572	12012	5148	8008	15912	182	2002	572	68068	6188
m_j	2	3	2/3	7/24	3/20	1	1	1/6	7/12	7/120

*For $t \geq 10$, the only values given are those for the first $t/2$ levels for even t or the first $(t+1)/2$ levels for odd t, i.e., for nonpositive values of $x-\bar{x}$. For even degrees, the values not shown for positive values of $x-\bar{x}$ form a mirror image of those given. For odd degrees they form a mirror image with opposite signs from those given.

Table A.7.1. Coefficients of Orthogonal Polynomials (ξ'_{ij}): Equally Spaced Levels (cont.)

i		t=14*					t=15*			
	j=1	2	3	4	5	j=1	2	3	4	5
1	-13	+13	-143	+143	-143	-7	+91	-91	+1001	-1001
2	-11	+7	-11	-77	+187	-6	+52	-13	-429	+1144
3	-9	+2	+66	-132	+132	-5	+19	+35	-869	+979
4	-7	-2	+98	-92	-28	-4	-8	+58	-704	+44
5	-5	-5	+95	-13	-139	-3	-29	+61	-249	-751
6	-3	-7	+67	+63	-145	-2	-44	+49	+251	-1000
7	-1	-8	+24	+108	-60	-1	-53	+27	+621	-675
8						0	-56	0	+756	0
$\sum_{i=1}^{t}(\xi'_{ij})^2$	910	728	97240	136136	235144	280	37128	39780	6466460	10581480
m_j	2	1/2	5/3	7/12	7/30	1	3	5/6	35/12	21/20

i		t=16*					t=17*			
	j=1	2	3	4	5	j=1	2	3	4	5
1	-15	+35	-455	+273	-143	-8	+40	-28	+52	-104
2	-13	+21	-91	-91	+143	-7	+25	-7	-13	+91
3	-11	+9	+143	-221	+143	-6	+12	+7	-39	+104
4	-9	-1	+267	-201	+33	-5	+1	+15	-39	+39
5	-7	-9	+301	-101	-77	-4	-8	+18	-24	-36
6	-5	-15	+265	+23	-131	-3	-15	+17	-3	-83
7	-3	-19	+179	+129	-115	-2	-20	+13	+17	-88
8	-1	-21	+63	+189	-45	-1	-23	+7	+31	-55
9						0	-24	0	+36	0
$\sum_{i=1}^{t}(\xi'_{ij})^2$	1360	5712	1007760	470288	201552	408	7752	3876	16796	100776
m_j	2	1	10/3	7/12	1/10	1	1	1/6	1/12	1/20

Source: Table A.7.1 was taken from Table 23 of Fisher and Yates, 1963, *Statistical tables for biological, agricultural and medical research*, 6th ed., published by Oliver & Boyd, Edinburgh, by permission of the authors and publishers.

*For $t \geq 10$, the only values given are those for the first $t/2$ levels for even t or the first $(t+1)/2$ levels for odd t, i.e., for nonpositive values of $x-\bar{x}$. For even degrees, the values not shown for positive values of $x-\bar{x}$ form a mirror image of those given. For odd degrees they form a mirror image with opposite signs from those given.

Table A.7.2. Coefficients of Orthogonal Polynomials (ξ'_{ij}): Unequally Spaced Levels* ($t = 3$)

i	Degree (j) 1	2	i	Degree (j) 1	2	i	Degree (j) 1	2
0	−4	+2	0	−5	+3	0	−2	+4
1	−1	−3	1	−2	−4	1	−1	−5
3	+5	+1	4	+7	+1	5	+3	+1
$\sum_{i=1}^{3}(\xi'_{ij})^2$	42	14		78	26		14	42
m_j	3	7/3		3	13/6		1	21/10
0	−7	+5	0	−8	+6	0	−3	+7
1	−4	−6	1	−5	−7	1	−2	−8
6	+11	+1	7	+13	+1	8	+5	+1
$\sum_{i=1}^{3}(\xi'_{ij})^2$	186	62		258	86		38	114
m_j	3	31/15		3	43/21		1	57/28
0	−10	+8	0	−11	+9	0	−4	+10
1	−7	−9	1	−8	−10	1	−3	−11
9	+17	+1	10	+19	+1	11	+7	+1
$\sum_{i=1}^{3}(\xi'_{ij})^2$	438	146		546	182		74	222
m_j	3	73/36		3	91/45		1	111/55
0	−7	+3	0	−3	+5	0	−11	+7
2	−1	−5	2	−1	−7	2	−5	−9
5	+8	+2	7	+4	+2	9	+16	+2
$\sum_{i=1}^{3}(\xi'_{ij})^2$	114	38		26	78		402	134
m_j	3	19/15		1	39/35		3	67/63
0	−10	+4	0	−11	+5	0	−13	+5
3	−1	−7	3	−2	−8	4	−1	−9
7	+11	+3	8	+13	+3	9	+14	+4
$\sum_{i=1}^{3}(\xi'_{ij})^2$	222	74		294	98		366	122
m_j	3	37/42		3	49/60		3	61/90

*Each set of coefficients may be used for levels with different origins (e.g., 0,1,3 for 1,2,4), for different widths of interval (e.g., 0,1,5 for 0,5,25), or both (e.g., 0,1,11 for 1,10,100). The relative proportions of the intervals are critical.

Table A.7.3. Coefficients of Orthogonal Polynomials (ξ'_{ij}): Unequally Spaced Levels* ($t = 4$)

		Degree (j)				Degree (j)	
i	1	2	3	i	1	2	3
0	-7	+7	-3	0	-2	+43	-6
1	-3	-4	+8	1	-1	-17	+15
2	+1	-8	-6	2	0	-49	-10
4	+9	+5	+1	5	+3	+23	+1
$\sum_{i=1}^{4}(\xi'_{ij})^2$	140	154	110		14	5068	362
m_j	4	7/2	55/12		1	14	181/30
0	-9	+172	-10	0	-2	+1	-1
1	-5	-52	+24	1	-1	-1	+2
2	-1	-193	-15	3	+1	-1	-2
6	+15	+73	+1	4	+2	+1	+1
$\sum_{i=1}^{4}(\xi'_{ij})^2$	332	74866	902		10	4	10
m_j	4	83/2	451/60		1	2/3	5/6
0	-9	+76	-8	0	-5	+9	-5
1	-5	-43	+15	1	-3	-3	+9
3	+3	-104	-10	3	+1	-13	-5
5	+11	+71	+3	6	+7	+7	+1
$\sum_{i=1}^{4}(\xi'_{ij})^2$	236	23482	398		84	308	132
m_j	4	59/2	199/60		2	7/3	22/15
0	-11	+20	-8	0	-3	+3	-9
1	-7	-4	+14	1	-2	-1	+14
3	+1	-29	-7	4	+1	-5	-7
7	+17	+13	+1	7	+4	+3	+2
$\sum_{i=1}^{4}(\xi'_{ij})^2$	460	1426	310		30	44	330
m_j	4	23/6	155/84		1	2/3	55/42
0	-7	+63	-10	0	-4	+107	-18
1	-5	-4	+15	1	-3	-5	+25
4	+1	-107	-6	5	+1	-205	-9
9	+11	+48	+1	10	+6	+103	+2
$\sum_{i=1}^{4}(\xi'_{ij})^2$	196	17738	362		62	64108	1034
m_j	2	49/6	181/180		1	62/5	517/450

*Each set of coefficients may be used for levels with different origins (e.g., 0,1,2,4 for 1,2,3,5), for different widths of interval (e.g., 0,1,3,6 for 0,5,15,30), or both (e.g., 0,1,3,7 for 15,20,30,50). The relative proportions of the intervals are critical.

TABLE A.8. UPPER PERCENTAGE POINTS $(q_{\alpha,t,\nu})$ OF STUDENTIZED RANGE (1-CDF)

ν	α	t=2	3	4	5	6	7	8	9	10
1	.20	4.353	6.615	8.075	9.138	9.966	10.64	11.21	11.70	12.12
	.10	8.929	13.44	16.36	18.49	20.15	21.51	22.64	23.62	24.48
	.05	17.97	26.98	32.82	37.08	40.41	43.12	45.40	47.36	49.07
	.01	90.03	135.0	164.3	185.6	202.2	215.8	227.2	237.0	245.6
2	.20	2.667	3.820	4.559	5.098	5.521	5.867	6.158	6.409	6.630
	.10	4.130	5.733	6.773	7.538	8.139	8.633	9.049	9.409	9.725
	.05	6.085	8.331	9.798	10.88	11.74	12.44	13.03	13.54	13.99
	.01	14.04	19.02	22.29	24.72	26.63	28.20	29.53	30.68	31.69
3	.20	2.316	3.245	3.833	4.261	4.597	4.872	5.104	5.305	5.481
	.10	3.328	4.467	5.199	5.738	6.162	6.511	6.806	7.062	7.287
	.05	4.501	5.910	6.825	7.502	8.037	8.478	8.853	9.177	9.462
	.01	8.261	10.62	12.17	13.33	14.24	15.00	15.64	16.20	16.69
4	.20	2.168	3.004	3.527	3.907	4.205	4.449	4.655	4.832	4.989
	.10	3.015	3.976	4.586	5.035	5.388	5.679	5.926	6.139	6.327
	.05	3.927	5.040	5.757	6.287	6.707	7.053	7.347	7.602	7.826
	.01	6.512	8.120	9.173	9.958	10.58	11.10	11.55	11.93	12.27
5	.20	2.087	2.872	3.358	3.712	3.988	4.214	4.405	4.570	4.715
	.10	2.850	3.717	4.264	4.664	4.979	5.238	5.458	5.648	5.816
	.05	3.635	4.602	5.218	5.673	6.033	6.330	6.582	6.802	6.995
	.01	5.702	6.976	7.804	8.421	8.913	9.321	9.669	9.972	10.24
6	.20	2.036	2.788	3.252	3.588	3.850	4.065	4.246	4.403	4.540
	.10	2.748	3.559	4.065	4.435	4.726	4.966	5.168	5.344	5.499
	.05	3.461	4.339	4.896	5.305	5.628	5.895	6.122	6.319	6.493
	.01	5.243	6.331	7.033	7.556	7.973	8.318	8.613	8.869	9.097
7	.20	2.001	2.731	3.179	3.503	3.756	3.962	4.136	4.287	4.419
	.10	2.680	3.451	3.931	4.280	4.555	4.780	4.972	5.137	5.283
	.05	3.344	4.165	4.681	5.060	5.359	5.606	5.815	5.998	6.158
	.01	4.949	5.919	6.543	7.005	7.373	7.679	7.939	8.166	8.368
8	.20	1.976	2.689	3.126	3.440	3.686	3.886	4.055	4.201	4.330
	.10	2.630	3.374	3.834	4.169	4.431	4.646	4.829	4.987	5.126
	.05	3.261	4.041	4.529	4.886	5.167	5.399	5.597	5.767	5.918
	.01	4.746	5.635	6.204	6.625	6.960	7.237	7.474	7.681	7.863
9	.20	1.956	2.658	3.085	3.393	3.633	3.828	3.994	4.136	4.261
	.10	2.592	3.316	3.761	4.084	4.337	4.545	4.721	4.873	5.007
	.05	3.199	3.949	4.415	4.756	5.024	5.244	5.432	5.595	5.739
	.01	4.596	5.428	5.957	6.348	6.658	6.915	7.134	7.325	7.495

TABLE A.8. UPPER PERCENTAGE POINTS $(q_{\alpha,t,\nu})$ OF STUDENTIZED RANGE (1-CDF) (cont.)

ν	α	t=2	3	4	5	6	7	8	9	10
10	.20	1.941	2.632	3.053	3.355	3.590	3.782	3.944	4.084	4.206
	.10	2.563	3.270	3.704	4.018	4.264	4.465	4.636	4.783	4.913
	.05	3.151	3.877	4.327	4.654	4.912	5.124	5.305	5.461	5.599
	.01	4.482	5.270	5.769	6.136	6.428	6.669	6.875	7.055	7.213
11	.20	1.928	2.612	3.027	3.325	3.557	3.745	3.905	4.042	4.162
	.10	2.540	3.234	3.658	3.965	4.205	4.401	4.568	4.711	4.838
	.05	3.113	3.820	4.256	4.574	4.823	5.028	5.202	5.353	5.487
	.01	4.392	5.146	5.621	5.970	6.247	6.476	6.672	6.842	6.992
12	.20	1.918	2.596	3.006	3.300	3.529	3.715	3.872	4.007	4.126
	.10	2.521	3.204	3.621	3.922	4.156	4.349	4.511	4.652	4.776
	.05	3.082	3.773	4.199	4.508	4.751	4.950	5.119	5.265	5.395
	.01	4.320	5.046	5.502	5.836	6.101	6.321	6.507	6.670	6.814
13	.20	1.910	2.582	2.988	3.279	3.505	3.689	3.844	3.978	4.095
	.10	2.505	3.179	3.589	3.885	4.116	4.305	4.464	4.602	4.724
	.05	3.055	3.735	4.151	4.453	4.690	4.885	5.049	5.192	5.318
	.01	4.260	4.964	5.404	5.727	5.981	6.192	6.372	6.528	6.667
14	.20	1.902	2.570	2.973	3.261	3.485	3.667	3.820	3.953	4.069
	.10	2.491	3.158	3.563	3.854	4.081	4.267	4.424	4.560	4.680
	.05	3.033	3.702	4.111	4.407	4.639	4.829	4.990	5.131	5.254
	.01	4.210	4.895	5.322	5.634	5.881	6.085	6.258	6.409	6.543
15	.20	1.896	2.560	2.960	3.246.	3.467	3.648	3.800	3.931	4.046
	.10	2.479	3.140	3.540	3.828	4.052	4.235	4.390	4.524	4.641
	.05	3.014	3.674	4.076	4.367	4.595	4.782	4.940	5.077	5.198
	.01	4.168	4.836	5.252	5.556	5.796	5.994	6.162	6.309	6.439
16	.20	1.891	2.551	2.948	3.232	3.452	3.631	3.782	3.912	4.026
	.10	2.469	3.124	3.520	3.804	4.026	4.207	4.360	4.492	4.608
	.05	2.998	3.649	4.046	4.333	4.557	4.741	4.897	5.031	5.150
	.01	4.131	4.786	5.192	5.489	5.722	5.915	6.079	6.222	6.349
17	.20	1.886	2.543	2.938	3.220	3.439	3.617	3.766	3.895	4.008
	.10	2.460	3.110	3.503	3.784	4.004	4.183	4.334	4.464	4.579
	.05	2.984	3.628	4.020	4.303	4.524	4.705	4.858	4.991	5.108
	.01	4.099	4.742	5.140	5.430	5.659	5.847	6.007	6.147	6.270
18	.20	1.882	2.536	2.930	3.210	3.427	3.604	3.753	3.881	3.993
	.10	2.452	3.098	3.488	3.767	3.984	4.161	4.311	4.440	4.554
	.05	2.971	3.609	3.997	4.277	4.495	4.673	4.824	4.956	5.071
	.01	4.071	4.703	5.094	5.379	5.603	5.788	5.944	6.081	6.201

TABLE A.8.   UPPER PERCENTAGE POINTS $(q_{\alpha,t,\nu})$ OF STUDENTIZED RANGE (1-CDF)
(<u>cont.</u>)

ν	α	t=2	3	4	5	6	7	8	9	10
19	.20	1.878	2.530	2.922	3.200	3.416	3.592	3.740	3.867	3.979
	.10	2.445	3.087	3.474	3.751	3.966	4.142	4.290	4.418	4.531
	.05	2.960	3.593	3.977	4.253	4.469	4.645	4.794	4.924	5.038
	.01	4.046	4.670	5.054	5.334	5.554	5.735	5.889	6.022	6.141
20	.20	1.874	2.524	2.914	3.192	3.407	3.582	3.729	3.855	3.966
	.10	2.439	3.078	3.462	3.736	3.950	4.124	4.271	4.398	4.510
	.05	2.950	3.578	3.958	4.232	4.445	4.620	4.768	4.896	5.008
	.01	4.024	4.639	5.018	5.294	5.510	5.688	5.839	5.970	6.087
24	.20	1.864	2.507	2.892	3.166	3.377	3.549	3.694	3.818	3.927
	.10	2.420	3.047	3.423	3.692	3.900	4.070	4.213	4.336	4.445
	.05	2.919	3.532	3.901	4.166	4.373	4.541	4.684	4.807	4.915
	.01	3.956	4.546	4.907	5.168	5.374	5.542	5.685	5.809	5.919
30	.20	1.853	2.490	2.870	3.140	3.348	3.517	3.659	3.781	3.887
	.10	2.400	3.017	3.386	3.648	3.851	4.016	4.155	4.275	4.381
	.05	2.888	3.486	3.845	4.102	4.302	4.464	4.602	4.720	4.824
	.01	3.889	4.455	4.799	5.048	5.242	5.401	5.536	5.653	5.756
40	.20	1.843	2.473	2.848	3.114	3.318	3.484	3.624	3.743	3.848
	.10	2.381	2.988	3.349	3.605	3.803	3.963	4.099	4.215	4.317
	.05	2.858	3.442	3.791	4.039	4.232	4.389	4.521	4.635	4.735
	.01	3.825	4.367	4.696	4.931	5.114	5.265	5.392	5.502	5.599
60	.20	1.833	2.456	2.826	3.089	3.290	3.452	3.589	3.707	3.809
	.10	2.363	2.959	3.312	3.562	3.755	3.911	4.042	4.155	4.254
	.05	2.829	3.399	3.737	3.977	4.163	4.314	4.441	4.550	4.646
	.01	3.762	4.282	4.595	4.818	4.991	5.133	5.253	5.356	5.447
120	.20	1.822	2.440	2.805	3.063	3.260	3.420	3.554	3.669	3.770
	.10	2.344	2.930	3.276	3.520	3.707	3.859	3.987	4.096	4.191
	.05	2.800	3.356	3.685	3.917	4.096	4.241	4.363	4.468	4.560
	.01	3.702	4.200	4.497	4.709	4.872	5.005	5.118	5.214	5.299
∞	.20	1.812	2.424	2.784	3.037	3.232	3.389	3.520	3.632	3.730
	.10	2.326	2.902	3.240	3.478	3.661	3.808	3.931	4.037	4.129
	.05	2.772	3.314	3.633	3.858	4.030	4.170	4.286	4.387	4.474
	.01	3.643	4.120	4.403	4.603	4.757	4.882	4.987	5.078	5.157

TABLE A.8. UPPER PERCENTAGE POINTS ($q_{\alpha,t,\nu}$) OF STUDENTIZED RANGE (1-CDF) (cont.)

ν	α	t=11	12	13	14	15	16	17	18	19
10	.20	4.316	4.414	4.503	4.585	4.660	4.730	4.795	4.856	4.913
	.10	5.029	5.134	5.229	5.317	5.397	5.472	5.542	5.607	5.668
	.05	5.722	5.833	5.935	6.028	6.114	6.194	6.269	6.339	6.405
	.01	7.356	7.485	7.603	7.712	7.812	7.906	7.993	8.076	8.153
11	.20	4.270	4.366	4.454	4.534	4.608	4.677	4.741	4.801	4.857
	.10	4.951	5.053	5.146	5.231	5.309	5.382	5.450	5.514	5.573
	.05	5.605	5.713	5.811	5.901	5.984	6.062	6.134	6.202	6.265
	.01	7.128	7.250	7.362	7.465	7.560	7.649	7.732	7.809	7.883
12	.20	4.231	4.327	4.413	4.492	4.565	4.633	4.696	4.755	4.810
	.10	4.886	4.986	5.077	5.160	5.236	5.308	5.374	5.436	5.495
	.05	5.511	5.615	5.710	5.798	5.878	5.953	6.023	6.089	6.151
	.01	6.943	7.060	7.167	7.265	7.356	7.441	7.520	7.594	7.665
13	.20	4.199	4.293	4.379	4.457	4.529	4.596	4.658	4.716	4.770
	.10	4.832	4.930	5.019	5.100	5.176	5.245	5.311	5.372	5.429
	.05	5.431	5.533	5.625	5.711	5.789	5.862	5.931	5.995	6.055
	.01	6.791	6.903	7.006	7.101	7.188	7.269	7.345	7.417	7.485
14	.20	4.172	4.265	4.349	4.426	4.498	4.564	4.625	4.683	4.737
	.10	4.786	4.882	4.970	5.050	5.124	5.192	5.256	5.316	5.373
	.05	5.364	5.463	5.554	5.637	5.714	5.786	5.852	5.915	5.974
	.01	6.664	6.772	6.871	6.962	7.047	7.126	7.199	7.268	7.333
15	.20	4.148	4.240	4.324	4.400	4.471	4.536	4.597	4.654	4.707
	.10	4.746	4.841	4.927	5.006	5.079	5.147	5.209	5.269	5.324
	.05	5.306	5.404	5.493	5.574	5.649	5.720	5.785	5.846	5.904
	.01	6.555	6.660	6.757	6.845	6.927	7.003	7.074	7.142	7.204
16	.20	4.127	4.218	4.301	4.377	4.447	4.512	4.572	4.628	4.681
	.10	4.712	4.805	4.890	4.968	5.040	5.107	5.169	5.227	5.282
	.05	5.256	5.352	5.439	5.520	5.593	5.662	5.727	5.786	5.843
	.01	6.462	6.564	6.658	6.744	6.823	6.898	6.967	7.032	7.093
17	.20	4.109	4.199	4.282	4.357	4.426	4.490	4.550	4.606	4.659
	.10	4.682	4.774	4.858	4.935	5.005	5.071	5.133	5.190	5.244
	.05	5.212	5.307	5.392	5.471	5.544	5.612	5.675	5.734	5.790
	.01	6.381	6.480	6.572	6.656	6.734	6.806	6.873	6.937	6.997
18	.20	4.093	4.182	4.264	4.339	4.407	4.471	4.531	4.586	4.638
	.10	4.655	4.746	4.829	4.905	4.975	5.040	5.101	5.158	5.211
	.05	5.174	5.267	5.352	5.429	5.501	5.568	5.630	5.688	5.743
	.01	6.310	6.407	6.497	6.579	6.655	6.725	6.792	6.854	6.912

TABLE A.8.   UPPER PERCENTAGE POINTS ($q_{\alpha,t,\nu}$) OF STUDENTIZED RANGE (1-CDF)
            (<u>cont.</u>)

ν	α	t=11	12	13	14	15	16	17	18	19
19	.20	4.078	4.167	4.248	4.323	4.391	4.454	4.513	4.569	4.620
	.10	4.631	4.721	4.803	4.879	4.948	5.012	5.073	5.129	5.182
	.05	5.140	5.231	5.315	5.391	5.462	5.528	5.589	5.647	5.701
	.01	6.247	6.342	6.430	6.510	6.585	6.654	6.719	6.780	6.837
20	.20	4.065	4.154	4.234	4.308	4.376	4.439	4.498	4.552	4.604
	.10	4.609	4.699	4.780	4.855	4.924	4.987	5.047	5.103	5.155
	.05	5.108	5.199	5.282	5.357	5.427	5.493	5.553	5.610	5.663
	.01	6.191	6.285	6.371	6.450	6.523	6.591	6.654	6.714	6.771
24	.20	4.024	4.111	4.190	4.262	4.329	4.391	4.448	4.502	4.552
	.10	4.541	4.628	4.708	4.780	4.847	4.909	4.966	5.021	5.071
	.05	5.012	5.099	5.179	5.251	5.319	5.381	5.439	5.494	5.545
	.01	6.017	6.106	6.186	6.261	6.330	6.394	6.453	6.510	6.563
30	.20	3.982	4.068	4.145	4.216	4.281	4.342	4.398	4.451	4.500
	.10	4.474	4.559	4.635	4.706	4.770	4.830	4.886	4.939	4.988
	.05	4.917	5.001	5.077	5.147	5.211	5.271	5.327	5.379	5.429
	.01	5.849	5.932	6.008	6.078	6.143	6.203	6.259	6.311	6.361
40	.20	3.941	4.025	4.101	4.170	4.234	4.293	4.348	4.399	4.447
	.10	4.408	4.490	4.564	4.632	4.695	4.752	4.807	4.857	4.905
	.05	4.824	4.904	4.977	5.044	5.106	5.163	5.216	5.266	5.313
	.01	5.686	5.764	5.835	5.900	5.961	6.017	6.069	6.119	6.165
60	.20	3.900	3.982	4.056	4.124	4.186	4.244	4.297	4.347	4.395
	.10	4.342	4.421	4.493	4.558	4.619	4.675	4.727	4.775	4.821
	.05	4.732	4.808	4.878	4.942	5.001	5.056	5.107	5.154	5.199
	.01	5.528	5.601	5.667	5.728	5.785	5.837	5.886	5.931	5.974
120	.20	3.859	3.938	4.011	4.077	4.138	4.194	4.246	4.295	4.341
	.10	4.276	4.353	4.422	4.485	4.543	4.597	4.647	4.694	4.738
	.05	4.641	4.714	4.781	4.842	4.898	4.950	4.998	5.044	5.086
	.01	5.375	5.443	5.505	5.562	5.614	5.662	5.708	5.750	5.790
∞	.20	3.817	3.895	3.966	4.030	4.089	4.144	4.195	4.242	4.287
	.10	4.211	4.285	4.351	4.412	4.468	4.519	4.568	4.612	4.654
	.05	4.552	4.622	4.685	4.743	4.796	4.845	4.891	4.934	4.974
	.01	5.227	5.290	5.348	5.400	5.448	5.493	5.535	5.574	5.611

TABLE A.8. UPPER PERCENTAGE POINTS $(q_{\alpha,t,\nu})$ OF STUDENTIZED RANGE (1-CDF)
(<u>cont.</u>)

ν	α	t=20	22	24	26	28	30	32	34	36
19	.20	4.669	4.759	4.840	4.914	4.981	5.044	5.102	5.156	5.206
	.10	5.232	5.324	5.407	5.483	5.552	5.616	5.676	5.732	5.784
	.05	5.752	5.846	5.932	6.009	6.081	6.147	6.209	6.267	6.321
	.01	6.891	6.992	7.082	7.166	7.242	7.313	7.379	7.440	7.498
20	.20	4.652	4.742	4.822	4.895	4.963	5.025	5.082	5.136	5.186
	.10	5.205	5.296	5.378	5.453	5.522	5.586	5.645	5.700	5.752
	.05	5.714	5.807	5.891	5.968	6.039	6.104	6.165	6.222	6.275
	.01	6.823	6.922	7.011	7.092	7.168	7.237	7.302	7.362	7.419
24	.20	4.599	4.687	4.766	4.838	4.904	4.964	5.021	5.073	5.122
	.10	5.119	5.208	5.287	5.360	5.427	5.489	5.546	5.600	5.650
	.05	5.594	5.683	5.764	5.838	5.906	5.968	6.027	6.081	6.132
	.01	6.612	6.705	6.789	6.865	6.936	7.001	7.062	7.119	7.173
30	.20	4.546	4.632	4.710	4.779	4.844	4.903	4.958	5.010	5.058
	.10	5.034	5.120	5.197	5.267	5.332	5.392	5.447	5.499	5.547
	.05	5.475	5.561	5.638	5.709	5.774	5.833	5.889	5.941	5.990
	.01	6.407	6.494	6.572	6.644	6.710	6.772	6.828	6.881	6.932
40	.20	4.493	4.576	4.652	4.720	4.783	4.841	4.895	4.945	4.993
	.10	4.949	5.032	5.107	5.174	5.236	5.294	5.347	5.397	5.444
	.05	5.358	5.439	5.513	5.581	5.642	5.700	5.753	5.803	5.849
	.01	6.209	6.289	6.362	6.429	6.490	6.547	6.600	6.650	6.697
60	.20	4.439	4.520	4.594	4.661	4.722	4.778	4.831	4.880	4.925
	.10	4.864	4.944	5.015	5.081	5.141	5.196	5.247	5.295	5.340
	.05	5.241	5.319	5.389	5.453	5.512	5.566	5.617	5.664	5.708
	.01	6.015	6.090	6.158	6.220	6.277	6.330	6.378	6.424	6.467
120	.20	4.384	4.463	4.535	4.600	4.659	4.714	4.765	4.812	4.857
	.10	4.779	4.856	4.924	4.987	5.044	5.097	5.146	5.192	5.235
	.05	5.126	5.200	5.266	5.327	5.382	5.434	5.481	5.526	5.568
	.01	5.827	5.897	5.959	6.016	6.069	6.117	6.162	6.204	6.244
∞	.20	4.329	4.405	4.475	4.537	4.595	4.648	4.697	4.743	4.786
	.10	4.694	4.767	4.832	4.892	4.947	4.997	5.044	5.087	5.128
	.05	5.012	5.081	5.144	5.201	5.253	5.301	5.346	5.388	5.427
	.01	5.645	5.709	5.766	5.818	5.866	5.911	5.952	5.990	6.026

TABLE A.8.   UPPER PERCENTAGE POINTS $(q_{\alpha,t,\nu})$ OF STUDENTIZED RANGE (1-CDF)
             (cont.)

ν	α	t=38	40	50	60	70	80	90	100
30	.20	5.103	5.146	5.329	5.475	5.597	5.701	5.791	5.871
	.10	5.593	5.636	5.821	5.969	6.093	6.198	6.291	6.372
	.05	6.037	6.080	6.267	6.417	6.543	6.650	6.744	6.827
	.01	6.978	7.023	7.215	7.370	7.500	7.611	7.709	7.796
40	.20	5.037	5.078	5.257	5.399	5.518	5.619	5.708	5.786
	.10	5.488	5.529	5.708	5.850	5.969	6.071	6.160	6.238
	.05	5.893	5.934	6.112	6.255	6.375	6.477	6.566	6.645
	.01	6.740	6.782	6.960	7.104	7.225	7.328	7.419	7.500
60	.20	4.969	5.009	5.183	5.321	5.437	5.535	5.621	5.697
	.10	5.382	5.422	5.593	5.730	5.844	5.941	6.026	6.102
	.05	5.750	5.789	5.958	6.093	6.206	6.303	6.387	6.462
	.01	6.507	6.546	6.710	6.843	6.954	7.050	7.133	7.207
120	.20	4.899	4.938	5.106	5.240	5.352	5.447	5.530	5.603
	.10	5.275	5.313	5.476	5.606	5.715	5.808	5.888	5.960
	.05	5.607	5.644	5.802	5.929	6.035	6.126	6.205	6.275
	.01	6.281	6.316	6.467	6.588	6.689	6.776	6.852	6.919
∞	.20	4.826	4.864	5.026	5.155	5.262	5.353	5.433	5.503
	.10	5.166	5.202	5.357	5.480	5.582	5.669	5.745	5.812
	.05	5.463	5.498	5.646	5.764	5.863	5.947	6.020	6.085
	.01	6.060	6.092	6.228	6.338	6.429	6.507	6.575	6.636

Source: Values were extracted by permission from H. L. Harter, 1969,
Order statistics and their use in testing and estimation, vol. 1, Aerosp. Res.
Lab. USAF (Washington: USGPO), pp. 648-57.

TABLE A.9.   UPPER PERCENTAGE POINTS OF DUNNETT's t DISTRIBUTION (1-CDF)

The tabulated values of $t_{D,\alpha,m,\nu}$ are valid for comparing each of the
means of m experimental groups with the mean of a control group when the vari-
ances of the means are homogeneous, $\sigma^2_{\bar{y}_i} = \sigma^2_{\bar{y}_c}$ $(i = 1,2,\ldots m)$, and replication
is balanced, $r_i = r_c$ $(i = 1,2,\ldots m)$.

In Table A.9.1 for two-sided comparisons, figures in parentheses following the critical values are adjustment factors to be used in the case of un-
equal variances of the means in the range $1 < \sigma^2_{\bar{y}_i}/\sigma^2_{\bar{y}_c} < 7$, or with homogeneous
variance within groups, for unbalanced replication with $1 < r_c/r_i < 7$. The
adjustment factor when multiplied by $1 - (\hat{\sigma}^2_{\bar{y}_c}/\hat{\sigma}^2_{\bar{y}_i})$ or $1 - (r_i/r_c)$, gives the
percentage increase required for the tabulated critical value.

Table A.9.1.   Upper Percentage Points of Dunnett's t Distribution: Two-Sided
               Comparisons with Control

m

ν	α	1	2	3	4	5	6	7
5	.05	2.57	3.03(2.3)	3.29(3.6)	3.48(4.6)	3.62(5.4)	3.73(5.9)	3.82(6.4)
	.01	4.03	4.63(1.8)	4.98(3.0)	5.22(3.9)	5.41(4.6)	5.56(5.2)	5.69(5.7)
6	.05	2.45	2.86(2.1)	3.10(3.4)	3.26(4.3)	3.39(5.0)	3.49(5.6)	3.57(6.0)
	.01	3.71	4.21(1.6)	4.51(2.7)	4.71(3.5)	4.87(4.1)	5.00(4.6)	5.10(5.1)
7	.05	2.36	2.75(2.0)	2.97(3.2)	3.12(4.1)	3.24(4.8)	3.33(5.3)	3.41(5.7)
	.01	3.50	3.95(1.5)	4.21(2.4)	4.39(3.1)	4.53(3.7)	4.64(4.2)	4.74(4.6)
8	.05	2.31	2.67(2.0)	2.88(3.1)	3.02(3.9)	3.13(4.5)	3.22(5.1)	3.29(5.5)
	.01	3.36	3.77(1.3)	4.00(2.2)	4.17(2.9)	4.29(3.4)	4.40(3.9)	4.48(4.2)
9	.05	2.26	2.61(1.9)	2.81(3.0)	2.95(3.8)	3.05(4.4)	3.14(4.9)	3.20(5.3)
	.01	3.25	3.63(1.2)	3.85(2.1)	4.01(2.7)	4.12(3.2)	4.22(3.6)	4.30(3.9)
10	.05	2.23	2.57(1.8)	2.76(2.9)	2.89(3.6)	2.99(4.2)	3.07(4.7)	3.14(5.1)
	.01	3.17	3.53(1.2)	3.74(1.9)	3.88(2.5)	3.99(3.0)	4.08(3.4)	4.16(3.7)
11	.05	2.20	2.53(1.8)	2.72(2.8)	2.84(3.5)	2.94(4.1)	3.02(4.6)	3.08(4.9)
	.01	3.11	3.45(1.1)	3.65(1.8)	3.79(2.4)	3.89(2.8)	3.98(3.2)	4.05(3.5)
12	.05	2.18	2.50(1.7)	2.68(2.7)	2.81(3.4)	2.90(4.0)	2.98(4.4)	3.04(4.8)
	.01	3.05	3.39(1.1)	3.58(1.7)	3.71(2.3)	3.81(2.7)	3.89(3.0)	3.96(3.3)
13	.05	2.16	2.48(1.7)	2.65(2.7)	2.78(3.4)	2.87(3.9)	2.94(4.3)	3.00(4.7)
	.01	3.01	3.33(1.0)	3.52(1.7)	3.65(2.2)	3.74(2.6)	3.82(2.9)	3.89(3.2)
14	.05	2.14	2.46(1.7)	2.63(2.6)	2.75(3.3)	2.84(3.8)	2.91(4.2)	2.97(4.6)
	.01	2.98	3.29(1.0)	3.47(1.6)	3.59(2.1)	3.69(2.5)	3.76(2.8)	3.83(3.0)
15	.05	2.13	2.44(1.7)	2.61(2.6)	2.73(3.2)	2.82(3.8)	2.89(4.2)	2.95(4.5)
	.01	2.95	3.25(0.9)	3.43(1.5)	3.55(2.0)	3.64(2.4)	3.71(2.7)	3.78(2.9)

Table A.9.1.   Upper Percentage Points of Dunnett's t Distribution: Two-Sided
                Comparisons with Control (<u>cont</u>.)

					m			
ν	α	1	2	3	4	5	6	7
16	.05	2.12	2.42(1.6)	2.59(2.5)	2.71(3.2)	2.80(3.7)	2.87(4.1)	2.29(4.4)
	.01	2.92	3.22(0.9)	3.39(1.5)	3.51(1.9)	3.60(2.3)	3.67(2.6)	3.73(2.8)
17	.05	2.11	2.41(1.6)	2.58(2.5)	2.69(3.1)	2.78(3.6)	2.85(4.0)	2.90(4.4)
	.01	2.90	3.19(0.9)	3.36(1.5)	3.47(1.9)	3.56(2.2)	3.63(2.5)	3.69(2.7)
18	.05	2.10	2.40(1.6)	2.56(2.5)	2.68(3.1)	2.76(3.6)	2.83(4.0)	2.89(4.3)
	.01	2.88	3.17(0.9)	3.33(1.4)	3.44(1.8)	3.53(2.2)	3.60(2.4)	3.66(2.7)
19	.05	2.09	2.39(1.6)	2.55(2.5)	2.66(3.1)	2.75(3.5)	2.81(3.9)	2.87(4.2)
	.01	2.86	3.15(0.9)	3.31(1.4)	3.42(1.8)	3.50(2.1)	3.57(2.4)	3.63(2.6)
20	.05	2.09	2.38(1.6)	2.54(2.4)	2.65(3.0)	2.73(3.5)	2.80(3.9)	2.86(4.2)
	.01	2.85	3.13(0.8)	3.29(1.4)	3.40(1.7)	3.48(2.1)	3.55(2.3)	3.60(2.5)
24	.05	2.06	2.35(1.5)	2.51(2.3)	2.61(2.9)	2.70(3.4)	2.76(3.7)	2.81(4.0)
	.01	2.80	3.07(0.8)	3.22(1.3)	3.32(1.6)	3.40(1.9)	3.47(2.1)	3.52(2.4)
30	.05	2.04	2.32(1.5)	2.47(2.3)	2.58(2.8)	2.66(3.2)	2.72(3.6)	2.77(3.9)
	.01	2.75	3.01(0.7)	3.15(1.2)	3.25(1.5)	3.33(1.8)	3.39(2.0)	3.44(2.2)
40	.05	2.02	2.29(1.4)	2.44(2.2)	2.54(2.7)	2.62(3.1)	2.68(3.4)	2.73(3.7)
	.01	2.70	2.95(0.7)	3.09(1.1)	3.19(1.4)	3.26(1.6)	3.32(1.8)	3.37(2.0)
60	.05	2.00	2.27(1.4)	2.41(2.1)	2.51(2.6)	2.58(3.0)	2.64(3.3)	2.69(3.5)
	.01	2.66	2.90(0.6)	3.03(1.0)	3.12(1.3)	3.19(1.5)	3.25(1.6)	3.29(1.8)
120	.05	1.98	2.24(1.3)	2.38(2.0)	2.47(2.5)	2.55(2.8)	2.60(3.1)	2.65(3.3)
	.01	2.62	2.85(0.6)	2.97(0.9)	3.06(1.1)	3.12(1.3)	3.18(1.5)	3.22(1.6)
∞	.05	1.96	2.21(1.3)	2.35(1.9)	2.44(2.3)	2.51(2.7)	2.57(2.9)	2.61(3.1)
	.01	2.58	2.79(0.5)	2.92(0.8)	3.00(1.0)	3.06(1.2)	3.11(1.3)	3.15(1.4)

Table A.9.1. Upper Percentage Points of Dunnett's t Distribution: Two-Sided
Comparisons with Control (<u>cont</u>.)

m

ν	α	8	9	10	11	12	15	20
16	.05	2.97(4.7)	3.02(5.0)	3.06(5.2)	3.09(5.4)	3.12(5.6)	3.20(6.1)	3.30(6.7)
	.01	3.78(3.1)	3.83(3.3)	3.87(3.4)	3.91(3.6)	3.94(3.8)	4.02(4.1)	4.13(4.6)
17	.05	2.95(4.7)	3.00(4.9)	3.03(5.1)	3.07(5.3)	3.10(5.5)	3.18(6.0)	3.27(6.6)
	.01	3.74(3.0)	3.79(3.2)	3.83(3.3)	3.86(3.5)	3.90(3.6)	3.98(4.0)	4.08(4.5)
18	.05	2.94(4.6)	2.98(4.8)	3.01(5.1)	3.05(5.3)	3.08(5.4)	3.16(5.9)	3.25(6.5)
	.01	3.71(2.9)	3.75(3.1)	3.79(3.2)	3.83(3.4)	3.86(3.5)	3.94(3.9)	4.04(4.4)
19	.05	2.92(4.5)	2.96(4.8)	3.00(5.0)	3.03(5.2)	3.06(5.4)	3.14(5.8)	3.23(6.4)
	.01	3.68(2.8)	3.72(3.0)	3.76(3.2)	3.79(3.3)	3.83(3.4)	3.90(3.8)	4.00(4.3)
20	.05	2.90(4.5)	2.95(4.7)	2.98(4.9)	3.02(5.1)	3.05(5.3)	3.12(5.7)	3.22(6.3)
	.01	3.65(2.7)	3.69(2.9)	3.73(3.1)	3.77(3.2)	3.80(3.4)	3.87(3.7)	3.97(4.2)
24	.05	2.86(4.3)	2.90(4.5)	2.94(4.7)	2.97(4.9)	3.00(5.1)	3.07(5.5)	3.16(6.0)
	.01	3.57(2.5)	3.61(2.7)	3.64(2.8)	3.68(3.0)	3.70(3.1)	3.78(3.4)	3.87(3.8)
30	.05	2.82(4.1)	2.86(4.3)	2.89(4.5)	2.92(4.7)	2.95(4.8)	3.02(5.2)	3.11(5.8)
	.01	3.49(2.3)	3.52(2.5)	3.56(2.6)	3.59(2.7)	3.62(2.8)	3.69(3.1)	3.78(3.5)
40	.05	2.77(3.9)	2.81(4.1)	2.85(4.3)	2.87(4.5)	2.90(4.6)	2.97(5.0)	3.06(5.5)
	.01	3.41(2.1)	3.44(2.3)	3.48(2.4)	3.51(2.5)	3.53(2.6)	3.60(2.8)	3.68(3.2)
60	.05	2.73(3.7)	2.77(3.9)	2.80(4.1)	2.83(4.2)	2.86(4.4)	2.92(4.7)	3.00(5.1)
	.01	3.33(1.9)	3.37(2.0)	3.40(2.1)	3.42(2.2)	3.45(2.3)	3.51(2.5)	3.59(2.8)
120	.05	2.69(3.5)	2.73(3.7)	2.76(3.8)	2.79(4.0)	2.81(4.1)	2.87(4.4)	2.95(4.8)
	.01	3.26(1.7)	3.29(1.8)	3.32(1.9)	3.35(2.0)	3.37(2.1)	3.43(2.2)	3.51(2.5)
∞	.05	2.65(3.3)	2.69(3.5)	2.72(3.6)	2.74(3.7)	2.77(3.8)	2.83(4.1)	2.91(4.5)
	.01	3.19(1.5)	3.22(1.6)	3.25(1.7)	3.27(1.7)	3.29(1.8)	3.35(1.9)	3.42(2.2)

Source: Values extracted by permission from C. W. Dunnett, *Biometrics* 20
(1964):488–89.

Table A.9.2. Upper Percentage Points of Dunnett's t Distribution: One-Sided
Comparisons with Control

m

ν	α	1	2	3	4	5	6	7	8	9
5	.05	2.02	2.44	2.68	2.85	2.98	3.08	3.16	3.24	3.30
	.01	3.37	3.90	4.21	4.43	4.60	4.73	4.85	4.94	5.03
6	.05	1.94	2.34	2.56	2.71	2.83	2.92	3.00	3.07	3.12
	.01	3.14	3.61	3.88	4.07	4.21	4.33	4.43	4.51	4.59
7	.05	1.89	2.27	2.48	2.62	2.73	2.82	2.89	2.95	3.01
	.01	3.00	3.42	3.66	3.83	3.96	4.07	4.15	4.23	4.30
8	.05	1.86	2.22	2.42	2.55	2.66	2.74	2.81	2.87	2.92
	.01	2.90	3.29	3.51	3.67	3.79	3.88	3.96	4.03	4.09
9	.05	1.83	2.18	2.37	2.50	2.60	2.68	2.75	2.81	2.86
	.01	2.82	3.19	3.40	3.55	3.66	3.75	3.82	3.89	3.94
10	.05	1.81	2.15	2.34	2.47	2.56	2.64	2.70	2.76	2.81
	.01	2.76	3.11	3.31	3.45	3.56	3.64	3.71	3.78	3.83
11	.05	1.80	2.13	2.31	2.44	2.53	2.60	2.67	2.72	2.77
	.01	2.72	3.06	3.25	3.38	3.48	3.56	3.63	3.69	3.74
12	.05	1.78	2.11	2.29	2.41	2.50	2.58	2.64	2.69	2.74
	.01	2.68	3.01	3.19	3.32	3.42	3.50	3.56	3.62	3.67
13	.05	1.77	2.09	2.27	2.39	2.48	2.55	2.61	2.66	2.71
	.01	2.65	2.97	3.15	3.27	3.37	3.44	3.51	3.56	3.61
14	.05	1.76	2.08	2.25	2.37	2.46	2.53	2.59	2.64	2.69
	.01	2.62	2.94	3.11	3.23	3.32	3.40	3.46	3.51	3.56
15	.05	1.75	2.07	2.24	2.36	2.44	2.51	2.57	2.62	2.67
	.01	2.60	2.91	3.08	3.20	3.29	3.36	3.42	3.47	3.52

Table A.9.2. Upper Percentage Points of Dunnett's t Distribution: One-Sided
              Comparisons with Control (<u>cont</u>.)

m

ν	α	1	2	3	4	5	6	7	8	9
16	.05	1.75	2.06	2.23	2.34	2.43	2.50	2.56	2.61	2.65
	.01	2.58	2.88	3.05	3.17	3.26	3.33	3.39	3.44	3.48
17	.05	1.74	2.05	2.22	2.33	2.42	2.49	2.54	2.59	2.64
	.01	2.57	2.86	3.03	3.14	3.23	3.30	3.36	3.41	3.45
18	.05	1.73	2.04	2.21	2.32	2.41	2.48	2.53	2.58	2.62
	.01	2.55	2.84	3.01	3.12	3.21	3.27	3.33	3.38	3.42
19	.05	1.73	2.03	2.20	2.31	2.40	2.47	2.52	2.57	2.61
	.01	2.54	2.83	2.99	3.10	3.18	3.25	3.31	3.36	3.40
20	.05	1.72	2.03	2.19	2.30	2.39	2.46	2.51	2.56	2.60
	.01	2.53	2.81	2.97	3.08	3.17	3.23	3.29	3.34	3.38
24	.05	1.71	2.01	2.17	2.28	2.36	2.43	2.48	2.53	2.57
	.01	2.49	2.77	2.92	3.03	3.11	3.17	3.22	3.27	3.31
30	.05	1.70	1.99	2.15	2.25	2.33	2.40	2.45	2.50	2.54
	.01	2.46	2.72	2.87	2.97	3.05	3.11	3.16	3.21	3.24
40	.05	1.68	1.97	2.13	2.23	2.31	2.37	2.42	2.47	2.51
	.01	2.42	2.68	2.82	2.92	2.99	3.05	3.10	3.14	3.18
60	.05	1.67	1.95	2.10	2.21	2.28	2.35	2.39	2.44	2.48
	.01	2.39	2.64	2.78	2.87	2.94	3.00	3.04	3.08	3.12
120	.05	1.66	1.93	2.08	2.18	2.26	2.32	2.37	2.41	2.45
	.01	2.36	2.60	2.73	2.82	2.89	2.94	2.99	3.03	3.06
∞	.05	1.64	1.92	2.06	2.16	2.23	2.29	2.34	2.38	2.42
	.01	2.33	2.56	2.68	2.77	2.84	2.89	2.93	2.97	3.00

Source: Values taken by permission from C. W. Dunnett, *J. Am. Stat.
Assoc. 50(1955):1115-16.

TABLE A.10.   UPPER PERCENTAGE POINTS FOR BONFERRONI t STATISTICS (1-CDF)

m	α	ν=10	12	14	16	18	20	22	24	26
1	.05	1.812	1.782	1.761	1.746	1.734	1.725	1.717	1.711	1.706
	.025	2.228	2.179	2.145	2.120	2.101	2.086	2.074	2.064	2.056
	.01	2.764	2.681	2.624	2.583	2.552	2.528	2.508	2.492	2.479
	.005	3.169	3.055	2.977	2.921	2.878	2.845	2.819	2.797	2.779
2	.05	2.228	2.179	2.145	2.120	2.101	2.086	2.074	2.064	2.056
	.025	2.634	2.560	2.510	2.473	2.445	2.423	2.405	2.391	2.379
	.01	3.169	3.055	2.977	2.921	2.878	2.845	2.819	2.797	2.779
	.005	3.581	3.428	3.326	3.252	3.197	3.158	3.119	3.091	3.067
3	.05	2.468	2.403	2.360	2.324	2.304	2.285	2.270	2.258	2.247
	.025	2.870	2.779	2.718	2.673	2.639	2.613	2.591	2.574	2.559
	.01	3.409	3.273	3.181	3.115	3.065	3.026	2.995	2.970	2.949
	.005	3.827	3.649	3.530	3.444	3.380	3.331	3.291	3.258	3.231
4	.05	2.634	2.560	2.510	2.473	2.445	2.428	2.405	2.391	2.379
	.025	3.038	2.934	2.864	2.813	2.775	2.744	2.720	2.700	2.684
	.01	3.581	3.428	3.326	3.252	3.197	3.153	3.119	3.091	3.067
	.005	4.005	3.807	3.675	3.581	3.510	3.455	3.412	3.376	3.346
5	.05	2.764	2.681	2.624	2.583	2.552	2.528	2.508	2.492	2.479
	.025	3.169	3.055	2.977	2.921	2.878	2.845	2.819	2.797	2.779
	.01	3.716	3.550	3.438	3.358	3.298	3.251	3.214	3.183	3.158
	.005	4.144	3.930	3.787	3.686	3.610	3.552	3.505	3.467	3.435
6	.05	2.870	2.779	2.718	2.673	2.639	2.613	2.591	2.574	2.559
	.025	3.277	3.153	3.069	3.008	2.963	2.927	2.899	2.875	2.856
	.01	3.827	3.649	3.530	3.444	3.380	3.331	3.291	3.258	3.231
	.005	4.259	4.031	3.880	3.773	3.692	3.630	3.581	3.540	3.507
7	.05	2.960	2.863	2.796	2.748	2.712	2.683	2.661	2.642	2.626
	.025	3.368	3.236	3.146	3.082	3.034	2.996	2.965	2.941	2.920
	.01	3.922	3.733	3.607	3.517	3.450	3.398	3.356	3.322	3.293
	.005	4.357	4.117	3.958	3.846	3.762	3.697	3.645	3.603	3.567
8	.05	3.038	2.934	2.864	2.813	2.775	2.744	2.720	2.700	2.684
	.025	3.448	3.308	3.214	3.146	3.095	3.055	3.023	2.997	2.975
	.01	4.005	3.807	3.675	3.581	3.510	3.455	3.412	3.376	3.346
	.005	4.442	4.192	4.026	3.909	3.822	3.754	3.700	3.656	3.620

TABLE A.10.   UPPER PERCENTAGE POINTS FOR BONFERRONI t STATISTICS (1-CDF)
(cont.)

m	α	ν=30	35	40	50	60	80	100	250	500
1	.05	1.697	1.690	1.684	1.676	1.671	1.664	1.660	1.651	1.648
	.025	2.042	2.030	2.021	2.009	2.000	1.990	1.984	1.970	1.965
	.01	2.457	2.438	2.423	2.403	2.390	2.374	2.364	2.341	2.334
	.005	2.750	2.724	2.704	2.678	2.660	2.639	2.626	2.596	2.586
2	.05	2.042	2.030	2.021	2.009	2.000	1.990	1.984	1.970	1.965
	.025	2.360	2.342	2.329	2.311	2.299	2.284	2.276	2.255	2.248
	.01	2.750	2.724	2.704	2.678	2.660	2.639	2.626	2.596	2.586
	.005	3.030	2.996	2.971	2.937	2.915	2.887	2.871	2.832	2.820
3	.05	2.231	2.215	2.204	2.188	2.178	2.165	2.158	2.140	2.134
	.025	2.536	2.515	2.499	2.477	2.463	2.445	2.435	2.410	2.402
	.01	2.915	2.885	2.862	2.831	2.811	2.786	2.771	2.736	2.725
	.005	3.189	3.150	3.122	3.083	3.057	3.026	3.007	2.964	2.949
4	.05	2.360	2.342	2.329	2.311	2.299	2.284	2.276	2.255	2.248
	.025	2.657	2.633	2.616	2.591	2.575	2.555	2.544	2.516	2.507
	.01	3.030	2.996	2.971	2.937	2.915	2.887	2.871	2.832	2.820
	.005	3.300	3.258	3.227	3.184	3.156	3.122	3.102	3.054	3.039
5	.05	2.457	2.438	2.423	2.403	2.390	2.374	2.364	2.341	2.334
	.025	2.750	2.724	2.704	2.678	2.660	2.639	2.626	2.596	2.586
	.01	3.118	3.081	3.055	3.018	2.994	2.964	2.945	2.905	2.892
	.005	3.385	3.340	3.307	3.261	3.232	3.195	3.174	3.123	3.107
6	.05	2.536	2.515	2.499	2.477	2.463	2.445	2.435	2.410	2.402
	.025	2.825	2.797	2.776	2.747	2.729	2.705	2.692	2.659	2.649
	.01	3.189	3.150	3.122	3.083	3.057	3.026	3.007	2.964	2.949
	.005	3.454	3.407	3.372	3.324	3.293	3.254	3.232	3.179	3.161
7	.05	2.601	2.579	2.562	2.539	2.524	2.505	2.494	2.467	2.459
	.025	2.887	2.857	2.836	2.805	2.785	2.761	2.747	2.712	2.701
	.01	3.249	3.208	3.178	3.137	3.111	3.078	3.058	3.013	2.998
	.005	3.513	3.463	3.426	3.376	3.344	3.304	3.280	3.225	3.207
8	.05	2.657	2.633	2.616	2.591	2.575	2.555	2.544	2.516	2.507
	.025	2.941	2.910	2.887	2.855	2.834	2.809	2.793	2.758	2.746
	.01	3.300	3.258	3.227	3.184	3.156	3.122	3.102	3.054	3.039
	.005	3.563	3.511	3.473	3.421	3.388	3.346	3.322	3.265	3.246

TABLE A.10.   UPPER PERCENTAGE POINTS FOR BONFERRONI t STATISTICS (1-CDF)
              (cont.)

m	α	ν=10	12	14	16	18	20	22	24	26
9	.05	3.107	2.998	2.924	2.870	2.829	2.798	2.772	2.751	2.734
	.025	3.518	3.371	3.273	3.202	3.149	3.107	3.074	3.046	3.024
	.01	4.078	3.871	3.734	3.636	3.563	3.506	3.461	3.424	3.393
	.005	4.518	4.258	4.086	3.965	3.874	3.804	3.749	3.703	3.666
10	.05	3.169	3.055	2.977	2.921	2.878	2.845	2.819	2.797	2.779
	.025	3.581	3.428	3.326	3.252	3.197	3.153	3.119	3.091	3.067
	.01	4.144	3.930	3.787	3.686	3.610	3.552	3.505	3.467	3.435
	.005	4.587	4.318	4.140	4.015	3.922	3.850	3.792	3.745	3.707
11	.05	3.225	3.106	3.025	2.967	2.923	2.888	2.861	2.838	2.819
	.025	3.639	3.480	3.374	3.297	3.240	3.195	3.159	3.130	3.106
	.01	4.204	3.982	3.836	3.731	3.653	3.593	3.545	3.505	3.473
	.005	4.649	4.372	4.189	4.060	3.964	3.890	3.831	3.783	3.744
12	.05	3.277	3.153	3.069	3.008	2.963	2.927	2.899	2.875	2.856
	.025	3.691	3.527	3.417	3.339	3.279	3.233	3.196	3.166	3.141
	.01	4.259	4.031	3.880	3.773	3.692	3.630	3.581	3.540	3.507
	.005	4.706	4.422	4.234	4.102	4.004	3.928	3.867	3.818	3.777
13	.05	3.324	3.196	3.109	3.047	3.000	2.963	2.933	2.909	2.889
	.025	3.740	3.571	3.458	3.377	3.316	3.268	3.230	3.199	3.174
	.01	4.309	4.075	3.921	3.810	3.728	3.665	3.614	3.573	3.538
	.005	4.759	4.467	4.275	4.140	4.039	3.962	3.900	3.850	3.808
14	.05	3.368	3.236	3.146	3.082	3.034	2.996	2.965	2.941	2.920
	.025	3.785	3.611	3.495	3.412	3.349	3.301	3.262	3.230	3.204
	.01	4.357	4.117	3.958	3.846	3.762	3.697	3.645	3.603	3.567
	.005	4.809	4.510	4.314	4.175	4.073	3.994	3.931	3.879	3.837
15	.05	3.409	3.273	3.181	3.115	3.065	3.026	2.995	2.970	2.949
	.025	3.827	3.649	3.530	3.444	3.380	3.331	3.291	3.258	3.231
	.01	4.401	4.156	3.993	3.878	3.793	3.726	3.673	3.630	3.594
	.005	4.855	4.550	4.349	4.208	4.104	4.023	3.959	3.907	3.864
16	.05	3.448	3.308	3.214	3.146	3.095	3.055	3.023	2.997	2.975
	.025	3.867	3.684	3.562	3.475	3.410	3.359	3.318	3.285	3.257
	.01	4.442	4.192	4.026	3.909	3.822	3.754	3.700	3.656	3.620
	.005	4.898	4.587	4.383	4.239	4.133	4.051	3.985	3.932	3.888

TABLE A.10.   UPPER PERCENTAGE POINTS FOR BONFERRONI t STATISTICS (1-CDF)
              (cont.)

m	α	ν=30	35	40	50	60	80	100	250	500
9	.05	2.706	2.681	2.663	2.637	2.620	2.600	2.587	2.558	2.549
	.025	2.988	2.955	2.931	2.898	2.877	2.850	2.834	2.797	2.785
	.01	3.345	3.301	3.269	3.225	3.196	3.161	3.140	3.091	3.075
	.005	3.607	3.553	3.514	3.461	3.426	3.383	3.358	3.299	3.280
10	.05	2.750	2.724	2.704	2.678	2.660	2.639	2.626	2.596	2.586
	.025	3.030	2.996	2.971	2.937	2.915	2.887	2.871	2.832	2.820
	.01	3.385	3.340	3.307	3.261	3.232	3.195	3.174	3.123	3.107
	.005	3.646	3.591	3.551	3.496	3.460	3.416	3.391	3.330	3.310
11	.05	2.789	2.762	2.742	2.714	2.696	2.674	2.660	2.629	2.619
	.025	3.067	3.033	3.007	2.972	2.948	2.920	2.903	2.864	2.851
	.01	3.421	3.375	3.341	3.294	3.264	3.226	3.204	3.152	3.135
	.005	3.681	3.625	3.584	3.528	3.491	3.446	3.420	3.358	3.338
12	.05	2.825	2.797	2.776	2.747	2.729	2.705	2.692	2.659	2.649
	.025	3.102	3.066	3.039	3.003	2.979	2.950	2.933	2.892	2.879
	.01	3.454	3.407	3.372	3.324	3.293	3.254	3.232	3.179	3.161
	.005	3.714	3.656	3.614	3.556	3.519	3.473	3.446	3.383	3.362
13	.05	2.857	2.828	2.807	2.778	2.758	2.734	2.720	2.687	2.676
	.025	3.133	3.096	3.069	3.032	3.007	2.977	2.960	2.918	2.904
	.01	3.485	3.436	3.400	3.351	3.319	3.280	3.257	3.203	3.185
	.005	3.743	3.685	3.642	3.583	3.545	3.498	3.470	3.406	3.385
14	.05	2.887	2.857	2.836	2.805	2.785	2.761	2.747	2.712	2.701
	.025	3.162	3.124	3.096	3.058	3.033	3.003	2.984	2.942	2.928
	.01	3.513	3.463	3.426	3.376	3.344	3.304	3.280	3.225	3.207
	.005	3.771	3.711	3.667	3.607	3.568	3.521	3.493	3.427	3.406
15	.05	2.915	2.885	2.862	2.831	2.811	2.786	2.771	2.736	2.725
	.025	3.189	3.150	3.122	3.083	3.057	3.026	3.007	2.964	2.949
	.01	3.538	3.488	3.451	3.400	3.366	3.326	3.302	3.245	3.227
	.005	3.796	3.735	3.691	3.630	3.590	3.542	3.513	3.446	3.425
16	.05	2.941	2.910	2.887	2.855	2.834	2.809	2.793	2.758	2.746
	.025	3.214	3.174	3.145	3.106	3.080	3.048	3.029	2.984	2.870
	.01	3.563	3.511	3.473	3.421	3.388	3.346	3.322	3.265	3.246
	.005	3.820	3.758	3.713	3.651	3.611	3.561	3.532	3.465	3.443

Source: Critical values extracted by permission from tables by C. M.
Dayton and W. D. Schafer, 1973 (Univ. Maryland, College Park, personal communication).

TABLE A.11.   UPPER PERCENTAGE POINTS FOR BONFERRONI CHI-SQUARE STATISTICS
              (1-CDF)

ν	α	m=1	2	3	4	5	6	7	8	9
1	.05	3.841	5.024	5.731	6.239	6.635	6.960	7.237	7.477	7.689
	.01	6.635	7.879	8.615	9.141	9.550	9.885	10.169	10.415	10.633
2	.05	5.991	7.378	8.189	8.764	9.210	9.575	9.883	10.150	10.386
	.01	9.210	10.597	11.408	11.983	12.429	12.794	13.102	13.369	13.605
3	.05	7.815	9.348	10.236	10.861	11.345	11.739	12.071	12.359	12.612
	.01	11.345	12.838	13.707	14.320	14.796	15.183	15.510	15.794	16.043
4	.05	9.488	11.143	12.094	12.762	13.277	13.695	14.048	14.353	14.621
	.01	13.277	14.860	15.777	16.424	16.924	17.331	17.675	17.972	18.233
5	.05	11.070	12.833	13.839	14.544	15.086	15.527	15.898	16.217	16.499
	.01	15.086	16.750	17.710	18.386	18.907	19.332	19.690	20.000	20.272
6	.05	12.592	14.449	15.506	16.245	16.812	17.272	17.659	17.993	18.286
	.01	16.812	18.548	19.547	20.249	20.791	21.232	21.603	21.924	22.206
7	.05	14.067	16.013	17.115	17.885	18.475	18.954	19.356	19.702	20.007
	.01	18.475	20.278	21.313	22.040	22.601	23.056	23.440	23.771	24.062
8	.05	15.507	17.535	18.680	19.478	20.090	20.586	21.002	21.360	21.675
	.01	20.090	21.955	23.024	23.774	24.352	24.821	25.216	25.557	25.857
9	.05	16.919	19.023	20.209	21.034	21.666	22.177	22.607	22.976	23.301
	.01	21.666	23.589	24.690	25.462	26.056	26.539	26.945	27.295	27.603
10	.05	18.307	20.483	21.707	22.558	23.209	23.736	24.178	24.558	24.891
	.01	23.209	25.188	26.320	27.112	27.722	28.216	28.633	28.991	29.307
11	.05	19.675	21.920	23.181	24.056	24.725	25.266	25.720	26.110	26.452
	.01	24.725	26.757	27.917	28.729	29.354	29.860	30.286	30.654	30.976
12	.05	21.026	23.337	24.632	25.530	26.217	26.772	27.237	27.637	27.987
	.01	26.217	28.300	29.487	30.318	30.957	31.475	31.910	32.286	32.615
13	.05	22.362	24.736	26.064	26.985	27.688	28.256	28.732	29.141	29.500
	.01	27.688	29.819	31.034	31.883	32.535	33.064	33.509	33.892	34.288

TABLE A.11. UPPER PERCENTAGE POINTS FOR BONFERRONI CHI-SQUARE STATISTICS
(1–CDF) (<u>cont.</u>)

ν	α	$m=1$	2	3	4	5	6	7	8	9
14	.05	23.685	26.119	27.480	28.422	29.141	29.722	30.209	30.627	30.993
	.01	29.141	31.319	32.559	33.426	34.091	34.631	35.084	35.475	35.818
15	.05	24.996	27.488	28.880	29.843	30.578	31.171	31.668	32.095	32.469
	.01	30.578	32.801	34.066	34.950	35.628	36.177	36.639	37.037	37.386
16	.05	26.296	28.845	30.267	31.250	32.000	32.605	33.111	33.547	33.928
	.01	32.000	34.267	35.556	36.456	37.146	37.706	38.175	38.580	38.936
17	.05	27.587	30.191	31.642	32.644	33.409	34.025	34.541	34.985	35.373
	.01	33.409	35.718	37.030	37.946	38.648	39.217	39.695	40.107	40.468
18	.05	28.869	31.526	33.005	34.027	34.805	35.433	35.959	36.410	36.805
	.01	34.805	37.156	38.491	39.422	40.136	40.714	41.200	41.618	41.985
19	.05	30.144	32.852	34.358	35.399	36.191	36.830	37.364	37.823	38.225
	.01	36.191	38.582	39.939	40.885	41.610	42.198	42.691	43.115	43.488
20	.05	31.410	34.170	35.702	36.761	37.566	38.216	38.759	39.225	39.634
	.01	37.566	39.997	41.375	42.336	43.072	43.668	44.169	44.600	44.978
21	.05	32.671	35.479	37.038	38.113	38.932	39.592	40.144	40.618	41.032
	.01	38.932	41.401	42.800	43.775	44.522	45.127	45.635	46.072	46.455
22	.05	33.924	36.781	38.365	39.458	40.289	40.959	41.520	42.001	42.422
	.01	40.289	42.796	44.215	45.204	45.962	46.575	47.090	47.533	47.922
23	.05	35.172	38.076	39.685	40.794	41.638	42.318	42.887	43.375	43.802
	.01	41.638	44.181	45.621	46.623	47.391	48.013	48.535	48.984	49.377
24	.05	36.415	39.364	40.998	42.123	42.980	43.669	44.246	44.741	45.174
	.01	42.980	45.559	47.018	48.034	48.812	49.441	49.970	50.424	50.823
25	.05	37.652	40.646	42.304	43.446	44.314	45.013	45.598	46.099	46.538
	.01	44.314	46.928	48.406	49.435	50.223	50.861	51.396	51.856	52.260

Source: Critical values extracted by permission from tables by C. M.
Dayton and W. D. Schafer, 1973 (Univ. Maryland, College Park, personal
communication). See C. Y. Kramer, 1972, *A first course in methods of multivariate analysis* (Va. Polytech. Inst. & State Univ., Dept. Stat., Blacksburg,
for $\alpha = 0.10$ or $m > 9$.

TABLE A.12.   RANKITS (Normal Order Statistics)*

Rank	n=1	2	3	4	5	6	7	8	9	10
1	0.00000	0.56419	0.84628	1.02938	1.16296	1.26721	1.35218	1.42360	1.48501	1.53875
2	---	---	0.00000	0.29701	0.49502	0.64176	0.75737	0.85222	0.93230	1.00136
3	---	---	---	---	0.00000	0.20155	0.35271	0.47282	0.57197	0.65606
4	---	---	---	---	---	---	0.00000	0.15251	0.27453	0.37576
5	---	---	---	---	---	---	---	---	0.00000	0.12267

Rank	n=11	12	13	14	15	16	17	18	19	20
1	1.58644	1.62923	1.66799	1.70338	1.73591	1.76599	1.79394	1.82003	1.84448	1.86748
2	1.06192	1.11573	1.16408	1.20790	1.24794	1.28474	1.31878	1.35041	1.37994	1.40760
3	0.72884	0.79284	0.84983	0.90113	0.94769	0.99027	1.02946	1.06573	1.09945	1.13095
4	0.46198	0.53684	0.60285	0.66176	0.71488	0.76317	0.80738	0.84812	0.88586	0.92098
5	0.22489	0.31225	0.38823	0.45557	0.51570	0.57001	0.61946	0.66479	0.70661	0.74538
6	0.00000	0.10259	0.19052	0.26730	0.33530	0.39622	0.45133	0.50158	0.54771	0.59030
7	---	---	0.00000	0.08816	0.16530	0.23375	0.29519	0.35084	0.40164	0.44833
8	---	---	---	---	0.00000	0.07729	0.14599	0.20774	0.26374	0.31493
9	---	---	---	---	---	---	0.00000	0.06880	0.13072	0.18696
10	---	---	---	---	---	---	---	---	0.00000	0.06200

Rank	n=21	22	23	24	25	26	27	28	29	30
1	1.88917	1.90969	1.92916	1.94767	1.96531	1.98216	1.99827	2.01371	2.02852	2.04276
2	1.43362	1.45816	1.48137	1.50338	1.52430	1.54423	1.56326	1.58145	1.59888	1.61560
3	1.16047	1.18824	1.21445	1.23924	1.26275	1.28511	1.30641	1.32674	1.34619	1.36481
4	0.95380	0.98459	1.01356	1.04091	1.06679	1.09135	1.11471	1.13697	1.15822	1.17855
5	0.78150	0.81527	0.84697	0.87682	0.9C501	0.93171	0.95705	0.98115	1.00414	1.02609
6	0.62982	0.66667	0.70115	0.73354	0.76405	0.79289	0.82021	0.84615	0.87084	0.89439
7	0.49148	0.53157	0.56896	0.60399	0.63690	0.66794	0.69727	0.72508	0.75150	0.77666
8	0.36203	0.40559	0.44609	0.48391	0.51935	0.55267	0.58411	0.61385	0.64205	0.66885
9	0.23841	0.28579	0.32965	0.37047	0.40860	0.44436	0.47801	0.50977	0.53982	0.56834
10	0.11836	0.16997	0.21755	0.26163	0.30268	0.34105	0.37706	0.41096	0.44298	0.47329
11	0.00000	0.05642	0.10813	0.15583	0.20006	0.24128	0.27983	0.31603	0.35013	0.38235
12	---	---	0.00000	0.05176	0.09953	0.14387	0.18520	0.22389	0.26023	0.29449
13	---	---	---	---	0.00000	0.04781	0.09220	0.13361	0.17240	0.20885
14	---	---	---	---	---	---	0.00000	0.04442	0.08588	0.12473
15	---	---	---	---	---	---	---	---	0.00000	0.04148

*This table lists expected values of ranked normal deviates for each position in the first half of the rank order (including the median for n odd) for samples of $n \leq 100$ observations. Values for the second half are symmetrical to those in the first half but have negative signs. Rankits may be used as substitutes for ranked data to permit use of standard analyses based on the normal distribution.

TABLE A.12.   RANKITS (Normal Order Statistics) (<u>cont</u>.)

Rank	n=31	32	33	34	35	36	37	38	39	40
1	2.05646	2.06967	2.08241	2.09471	2.10661	2.11812	2.12928	2.14009	2.15059	2.16078
2	1.63166	1.64712	1.66200	1.67636	1.69023	1.70362	1.71659	1.72914	1.74131	1.75312
3	1.38268	1.39985	1.41637	1.43228	1.44762	1.46244	1.47676	1.49061	1.50402	1.51702
4	1.19803	1.21672	1.23468	1.25196	1.26860	1.28466	1.30016	1.31514	1.32964	1.34368
5	1.04709	1.06721	1.08652	1.10509	1.12295	1.14016	1.15677	1.17280	1.18830	1.20330
6	0.91688	0.93841	0.95905	0.97886	0.99790	1.01624	1.03390	1.05095	1.06741	1.08332
7	0.80066	0.82359	0.84555	0.86660	0.88681	0.90625	0.92496	0.94300	0.96041	0.97722
8	0.69438	0.71875	0.74204	0.76435	0.78574	0.80629	0.82605	0.84508	0.86343	0.88114
9	0.59545	0.62129	0.64596	0.66954	0.69214	0.71382	0.73465	0.75468	0.77398	0.79259
10	0.50206	0.52943	0.55552	0.58043	0.60427	0.62710	0.64902	0.67009	0.69035	0.70988
11	0.41287	0.44185	0.46942	0.49572	0.52084	0.54488	0.56793	0.59005	0.61131	0.63177
12	0.32686	0.35755	0.38669	0.41444	0.44091	0.46620	0.49042	0.51363	0.53592	0.55736
13	0.24322	0.27573	0.30654	0.33582	0.36371	0.39032	0.41576	0.44012	0.46348	0.48591
14	0.16126	0.19572	0.22832	0.25924	0.28863	0.31663	0.34336	0.36892	0.39340	0.41688
15	0.08037	0.11695	0.15147	0.18415	0.21515	0.24463	0.27272	0.29954	0.32520	0.34978
16	0.00000	0.03890	0.07552	0.11009	0.14282	0.17388	0.20342	0.23159	0.25849	0.28423
17	---	---	0.00000	0.03663	0.07123	0.10399	0.13509	0.16469	0.19292	0.21988
18	---	---	---	---	0.00000	0.03461	0.06739	0.09853	0.12817	0.15644
19	---	---	---	---	---	---	0.00000	0.03280	0.06395	0.09362
20	---	---	---	---	---	---	---	---	0.00000	0.03117

Rank	n=41	42	43	44	45	46	47	48	49	50
1	2.17068	2.18032	2.18969	2.19882	2.20772	2.21639	2.22486	2.23312	2.24119	2.24907
2	1.76458	1.77571	1.78654	1.79707	1.80733	1.81732	1.82706	1.83655	1.84582	1.85487
3	1.52964	1.54188	1.55377	1.56533	1.57658	1.58754	1.59820	1.60860	1.61874	1.62863
4	1.35728	1.37048	1.38329	1.39574	1.40784	1.41962	1.43108	1.44224	1.45312	1.46374
5	1.21782	1.23190	1.24556	1.25881	1.27170	1.28422	1.29641	1.30827	1.31983	1.33109
6	1.09872	1.11364	1.12810	1.14213	1.15576	1.16899	1.18186	1.19439	1.20658	1.21846
7	0.99348	1.00922	1.02446	1.03924	1.05358	1.06751	1.08014	1.09420	1.10701	1.11948
8	0.89825	0.91480	0.93082	0.94634	0.96139	0.97599	0.99018	1.00396	1.01737	1.03042
9	0.81056	0.82792	0.84472	0.86097	0.87673	0.89201	0.90684	0.92125	0.93525	0.94887
10	0.72871	0.74690	0.76448	0.78148	0.79795	0.81391	0.82939	0.84442	0.85902	0.87321
11	0.65149	0.67052	0.68889	0.70666	0.72385	0.74049	0.75663	0.77228	0.78748	0.80225
12	0.57799	0.59788	0.61707	0.63561	0.65353	0.67088	0.68768	0.70397	0.71978	0.73513
13	0.50749	0.52827	0.54803	0.56763	0.58631	0.60438	0.62186	0.63881	0.65523	0.67117
14	0.43944	0.46114	0.48204	0.50220	0.52166	0.54046	0.55865	0.57625	0.59331	0.60986
15	0.37337	0.39604	0.41784	0.43885	0.45912	0.47868	0.49759	0.51588	0.53360	0.55077
16	0.30890	0.33257	0.35533	0.37723	0.39833	0.41868	0.43834	0.45734	0.47573	0.49354
17	0.24569	0.27043	0.29418	0.31701	0.33898	0.36016	0.38060	0.40034	0.41942	0.43789
18	0.18345	0.20931	0.23411	0.25792	0.28081	0.30285	0.32410	0.34460	0.36441	0.38357
19	0.12192	0.14897	0.17488	0.19972	0.22358	0.24652	0.26862	0.28992	0.31049	0.38036
20	0.06085	0.08917	0.11625	0.14219	0.16707	0.19097	0.21396	0.23610	0.25746	0.27807
21	0.00000	0.02969	0.05803	0.08513	0.11109	0.13600	0.15993	0.18296	0.20514	0.22653
22	---	---	0.00000	0.02835	0.05546	0.08144	0.10637	0.13033	0.15338	0.17559
23	---	---	---	---	0.00000	0.02712	0.05311	0.07805	0.10203	0.12511
24	---	---	---	---	---	---	0.00000	0.02599	0.05095	0.07494
25	---	---	---	---	---	---	---	---	0.00000	0.02496

TABLE A.12. RANKITS (Normal Order Statistics) (<u>cont.</u>)

Rank	n=51	52	53	54	55	56	57	58	59	60
1	2.25678	2.26432	2.27169	2.27891	2.28598	2.29291	2.29970	2.30635	2.31288	2.31928
2	1.86371	1.87235	1.88080	1.88906	1.89715	1.90506	1.91282	1.92041	1.92786	1.93516
3	1.63829	1.64773	1.65695	1.66596	1.67478	1.68340	1.69185	1.70012	1.70822	1.71616
4	1.47409	1.48420	1.49407	1.50372	1.51315	1.52237	1.53140	1.54024	1.54889	1.55736
5	1.34207	1.35279	1.36326	1.37348	1.38346	1.39323	1.40278	1.41212	1.42127	1.43023
6	1.23003	1.24132	1.25234	1.26310	1.27361	1.28387	1.29391	1.30373	1.31334	1.32274
7	1.13162	1.14347	1.15502	1.16629	1.17729	1.18804	1.19855	1.20882	1.21886	1.22869
8	1.04312	1.05550	1.06757	1.07934	1.09083	1.10205	1.11300	1.12371	1.13419	1.14443
9	0.96213	0.97504	0.98762	0.99988	1.01185	1.02352	1.03493	1.04607	1.05695	1.06760
10	0.88701	0.90045	0.91354	0.92629	0.93873	0.95086	0.96271	0.97427	0.98557	0.99662
11	0.81661	0.83058	0.84417	0.85742	0.87033	0.88292	0.89520	0.90719	0.91890	0.93034
12	0.75004	0.76455	0.77866	0.79240	0.80578	0.81883	0.83155	0.84397	0.85609	0.86793
13	0.68666	0.70170	0.71633	0.73057	0.74444	0.75794	0.77111	0.78396	0.79649	0.80873
14	0.62592	0.64152	0.65668	0.67143	0.68578	0.69976	0.71337	0.72665	0.73960	0.75224
15	0.56742	0.58358	0.59928	0.61455	0.62940	0.64385	0.65793	0.67164	0.68502	0.69807
16	0.51080	0.52755	0.54380	0.55960	0.57495	0.58989	0.60444	0.61860	0.63241	0.64587
17	0.45578	0.47312	0.48995	0.50629	0.52217	0.53761	0.55263	0.56725	0.58150	0.59538
18	0.40211	0.42007	0.43749	0.45439	0.47080	0.48675	0.50226	0.51736	0.53205	0.54637
19	0.34957	0.36818	0.38621	0.40369	0.42065	0.43713	0.45314	0.46872	0.48388	0.49864
20	0.29799	0.31726	0.33592	0.35400	0.37154	0.38856	0.40510	0.42117	0.43681	0.45202
21	0.24719	0.26716	0.28648	0.30518	0.32331	0.34090	0.35797	0.37456	0.39068	0.40637
22	0.19702	0.21772	0.23772	0.25708	0.27538	0.29400	0.31163	0.32875	0.34538	0.36155
23	0.14735	0.16880	0.18953	0.20957	0.22896	0.24774	0.26595	0.28362	0.30078	0.31745
24	0.09803	0.12029	0.14177	0.16252	0.18259	0.20201	0.22082	0.23906	0.25677	0.27396
25	0.04896	0.07206	0.09434	0.11584	0.13661	0.15669	0.17614	0.19498	0.21325	0.23098
26	0.00000	0.02400	0.04712	0.06940	0.09091	0.11170	0.13180	0.15127	0.17013	0.18842
27	---	---	0.00000	0.02312	0.04541	0.06693	0.08773	0.10785	0.12733	0.14621
28	---	---	---	---	0.00000	0.02229	0.04382	0.06463	0.08476	0.10425
29	---	---	---	---	---	---	0.00000	0.02153	0.04234	0.06248
30	---	---	---	---	---	---	---	---	0.00000	0.02081

TABLE A.12.   RANKITS (Normal Order Statistics) (<u>cont</u>.)

Rank	n=61	62	63	64	65	66	67	68	69	70
1	2.32556	2.33173	2.33778	2.34373	2.34958	2.35532	2.36097	2.36652	2.37199	2.37736
2	1.94232	1.94934	1.95624	1.96301	1.96965	1.97618	1.98260	1.98891	1.99510	2.00120
3	1.72394	1.73158	1.73906	1.74641	1.75363	1.76071	1.76767	1.77451	1.78122	1.78783
4	1.56567	1.57381	1.58180	1.58963	1.59732	1.60487	1.61228	1.61955	1.62670	1.63373
5	1.43900	1.44760	1.45603	1.46430	1.47241	1.48036	1.48817	1.49584	1.50338	1.51078
6	1.33195	1.34097	1.34982	1.35848	1.36698	1.37532	1.38351	1.39154	1.39942	1.40717
7	1.23832	1.24774	1.25698	1.26603	1.27490	1.28360	1.29213	1.30051	1.30873	1.31680
8	1.15445	1.16427	1.17388	1.18329	1.19252	1.20157	1.21044	1.21915	1.22769	1.23608
9	1.07802	1.08821	1.09819	1.10797	1.11754	1.12693	1.13613	1.14516	1.15401	1.16270
10	1.00742	1.01799	1.02833	1.03846	1.04838	1.05810	1.06762	1.07696	1.08612	1.09511
11	0.94153	0.95247	0.96317	0.97365	0.98391	0.99395	1.00380	1.01345	1.02291	1.03220
12	0.87950	0.89081	0.90187	0.91270	0.92329	0.93367	0.94383	0.95379	0.96355	0.97313
13	0.82068	0.83237	0.84379	0.85496	0.86590	0.87660	0.88708	0.89735	0.90741	0.91728
14	0.76459	0.77665	0.78843	0.79996	0.81123	0.82226	0.83306	0.84364	0.85400	0.86416
15	0.71081	0.72324	0.73540	0.74727	0.75889	0.77025	0.78138	0.79226	0.80293	0.81338
16	0.65901	0.67183	0.68436	0.69659	0.70856	0.72025	0.73170	0.74290	0.75387	0.76462
17	0.60893	0.62214	0.63504	0.64764	0.65996	0.67200	0.68377	0.69529	0.70657	0.71761
18	0.56033	0.57395	0.58723	0.60020	0.61288	0.62526	0.63737	0.64921	0.66080	0.67214
19	0.51303	0.52705	0.54073	0.55408	0.56712	0.57985	0.59230	0.60447	0.61638	0.62803
20	0.46685	0.48129	0.49537	0.50911	0.52252	0.53561	0.54841	0.56091	0.57314	0.58510
21	0.42164	0.43652	0.45101	0.46515	0.47894	0.49240	0.50555	0.51839	0.53095	0.54323
22	0.37729	0.39260	0.40752	0.42207	0.43625	0.45009	0.46360	0.47680	0.48969	0.50230
23	0.33366	0.34944	0.36480	0.37976	0.39435	0.40857	0.42245	0.43601	0.44925	0.46219
24	0.29066	0.30691	0.32272	0.33812	0.35312	0.36775	0.38201	0.39594	0.40953	0.42281
25	0.24820	0.26494	0.28122	0.29706	0.31249	0.32753	0.34219	0.35649	0.37045	0.38404
26	0.20618	0.22343	0.24019	0.25650	0.27237	0.28784	0.30290	0.31759	0.33192	0.34591
27	0.16452	0.18230	0.19957	0.21636	0.23269	0.24859	0.26408	0.27917	0.29389	0.30825
28	0.12315	0.14148	0.15927	0.17656	0.19337	0.20973	0.22565	0.24116	0.25627	0.27102
29	0.08198	0.10089	0.11923	0.13704	0.15435	0.17118	0.18755	0.20349	0.21902	0.23416
30	0.04096	0.06047	0.07938	0.09774	0.11556	0.13288	0.14972	0.16611	0.18207	0.19762
31	0.00000	0.02014	0.03966	0.05858	0.07694	0.09478	0.11211	0.12896	0.14536	0.16134
32	---	---	0.00000	0.01952	0.03844	0.05681	0.07465	0.09199	0.10885	0.12527
33	---	---	---	---	0.00000	0.01893	0.03730	0.05514	0.07249	0.08936
34	---	---	---	---	---	---	0.00000	0.01837	0.03622	0.05357
35	---	---	---	---	---	---	---	---	0.00000	0.01785

TABLE A.12. RANKITS (Normal Order Statistics) (<u>cont</u>.)

Rank	n=71	72	73	74	75	76	77	78	79	80
1	2.38265	2.38785	2.39298	2.39802	2.40299	2.40789	2.41271	2.41747	2.42215	2.42677
2	2.00720	2.01310	2.01890	2.02462	2.03024	2.03578	2.04124	2.04662	2.05191	2.05714
3	1.79432	1.80071	1.80699	1.81317	1.81926	1.82525	1.83115	1.83696	1.84268	1.84832
4	1.64063	1.64742	1.65410	1.66067	1.66714	1.67350	1.67976	1.68592	1.69200	1.69798
5	1.51805	1.52520	1.53223	1.53914	1.54594	1.55263	1.55921	1.56569	1.57207	1.57836
6	1.41478	1.42226	1.42961	1.43684	1.44395	1.45094	1.45782	1.46459	1.47125	1.47781
7	1.32473	1.33252	1.34017	1.34770	1.35510	1.36237	1.36953	1.37657	1.38350	1.39032
8	1.24431	1.25240	1.26034	1.26815	1.27583	1.28338	1.29080	1.29810	1.30529	1.31236
9	1.17123	1.17961	1.18784	1.19592	1.20387	1.21168	1.21936	1.22691	1.23434	1.24165
10	1.10393	1.11259	1.12110	1.12945	1.13766	1.14572	1.15365	1.16145	1.16912	1.17666
11	1.04130	1.05024	1.05902	1.06764	1.07610	1.08442	1.09260	1.10063	1.10854	1.11631
12	0.98252	0.99173	1.00078	1.00966	1.01838	1.02695	1.03537	1.04364	1.05178	1.05978
13	0.92695	0.93644	0.94576	0.95490	0.96387	0.97269	0.98135	0.98986	0.99822	1.00644
14	0.87412	0.88388	0.89346	0.90286	0.91209	0.92115	0.93005	0.93880	0.94739	0.95584
15	0.82362	0.83366	0.84351	0.85317	0.86265	0.87196	0.88110	0.89008	0.89890	0.90757
16	0.77514	0.78546	0.79558	0.80550	0.81524	0.82480	0.83418	0.84339	0.85244	0.86134
17	0.72843	0.73903	0.74942	0.75960	0.76960	0.77940	0.78903	0.79848	0.80776	0.81687
18	0.68325	0.69413	0.70480	0.71526	0.72551	0.73557	0.74544	0.75512	0.76463	0.77398
19	0.63943	0.65060	0.66155	0.67227	0.68279	0.69310	0.70322	0.71314	0.72289	0.73246
20	0.59681	0.60827	0.61950	0.63050	0.64128	0.65185	0.66222	0.67239	0.68237	0.69217
21	0.55525	0.56701	0.57842	0.58980	0.60085	0.61168	0.62230	0.63272	0.64294	0.65297
22	0.51463	0.52669	0.53850	0.55006	0.56138	0.57248	0.58336	0.59403	0.60449	0.61476
23	0.47484	0.48721	0.49932	0.51117	0.52277	0.53414	0.54528	0.55621	0.56692	0.57742
24	0.43579	0.44848	0.46089	0.47304	0.48493	0.49657	0.50798	0.51917	0.53013	0.54088
25	0.39739	0.41041	0.42313	0.43558	0.44777	0.45970	0.47138	0.48283	0.49404	0.50504
26	0.35958	0.37292	0.38597	0.39873	0.41122	0.42343	0.43540	0.44711	0.45859	0.46985
27	0.32227	0.33596	0.34934	0.36242	0.37521	0.38772	0.39997	0.41196	0.42371	0.43522
28	0.28540	0.29945	0.31317	0.32657	0.33968	0.35250	0.36504	0.37731	0.38934	0.40111
29	0.24893	0.26333	0.27740	0.29114	0.30457	0.31770	0.33055	0.34311	0.35542	0.36747
30	0.21277	0.22756	0.24199	0.25608	0.26984	0.28329	0.29645	0.30931	0.32190	0.33423
31	0.17690	0.19208	0.20688	0.22133	0.23543	0.24922	0.26269	0.27586	0.28875	0.30136
32	0.14125	0.15683	0.17202	0.18684	0.20130	0.21543	0.22923	0.24272	0.25591	0.26881
33	0.10579	0.12178	0.13737	0.15257	0.16740	0.18188	0.19602	0.20983	0.22334	0.23655
34	0.07045	0.08688	0.10289	0.11848	0.13370	0.14854	0.16303	0.17718	0.19101	0.20453
35	0.03520	0.05209	0.06852	0.08453	0.10014	0.11536	0.13021	0.14471	0.15888	0.17272
36	0.00000	0.01736	0.03424	0.05068	0.06670	0.08231	0.09754	0.11240	0.12691	0.14108
37	---	---	0.00000	0.01689	0.03333	0.04935	0.06497	0.08020	0.09507	0.10959
38	---	---	---	---	0.00000	0.01644	0.03247	0.04809	0.06333	0.07820
39	---	---	---	---	---	---	0.00000	0.01602	0.03165	0.04689
40	---	---	---	---	---	---	---	---	0.00000	0.01562

TABLE A.12.   RANKITS (Normal Order Statistics) (<u>cont</u>.)

Rank	n=81	82	83	84	85	86	87	88	89	90
1	2.43133	2.43582	2.44026	2.44463	2.44894	2.45320	2.45741	2.46156	2.46565	2.46970
2	2.06228	2.06735	2.07236	2.07729	2.08216	2.08696	2.09170	2.09637	2.10099	2.10554
3	1.85387	1.85935	1.86475	1.87007	1.87532	1.88049	1.88560	1.89064	1.89561	1.90052
4	1.70387	1.70968	1.71540	1.72104	1.72660	1.73209	1.73750	1.74283	1.74810	1.75329
5	1.58455	1.59065	1.59665	1.60258	1.60841	1.61417	1.61984	1.62544	1.63096	1.63641
6	1.48428	1.49064	1.49691	1.50309	1.50918	1.51518	1.52110	1.52693	1.53269	1.53836
7	1.39704	1.40366	1.41017	1.41659	1.42292	1.42915	1.43529	1.44135	1.44732	1.45321
8	1.31932	1.32617	1.33292	1.33957	1.34611	1.35257	1.35893	1.36520	1.37138	1.37747
9	1.24884	1.25593	1.26290	1.26977	1.27653	1.28320	1.28976	1.29624	1.30262	1.30891
10	1.18409	1.19139	1.19859	1.20567	1.21264	1.21951	1.22628	1.23295	1.23952	1.24600
11	1.12396	1.13148	1.13889	1.14618	1.15336	1.16043	1.16740	1.17426	1.18102	1.18769
12	1.06764	1.07539	1.08300	1.09050	1.09788	1.10515	1.11231	1.11936	1.12631	1.13316
13	1.01453	1.02249	1.03031	1.03802	1.04560	1.05306	1.06041	1.06765	1.07478	1.08181
14	0.96414	0.97231	0.98034	0.98825	0.99603	1.00369	1.01122	1.01865	1.02596	1.03316
15	0.91609	0.92447	0.93271	0.94082	0.94880	0.95665	0.96437	0.97198	0.97948	0.98686
16	0.87007	0.87867	0.88711	0.89542	0.90360	0.91164	0.91956	0.92735	0.93502	0.94258
17	0.82583	0.83464	0.84329	0.85180	0.86017	0.86841	0.87651	0.88449	0.89234	0.90007
18	0.78315	0.79217	0.80103	0.80975	0.81832	0.82675	0.83504	0.84320	0.85123	0.85914
19	0.74186	0.75109	0.76016	0.76908	0.77785	0.78647	0.79496	0.80330	0.81152	0.81960
20	0.70179	0.71124	0.72053	0.72965	0.73862	0.74744	0.75611	0.76465	0.77304	0.78131
21	0.66282	0.67249	0.68199	0.69133	0.70050	0.70952	0.71838	0.72710	0.73568	0.74412
22	0.62484	0.63473	0.64445	0.65399	0.66337	0.67259	0.68165	0.69056	0.69932	0.70795
23	0.58773	0.59785	0.60779	0.61755	0.62714	0.63656	0.64581	0.65492	0.66387	0.67267
24	0.55143	0.56178	0.57193	0.58191	0.59171	0.60133	0.61079	0.62009	0.62923	0.63822
25	0.51583	0.52641	0.53680	0.54700	0.55701	0.56684	0.57650	0.58600	0.59533	0.60451
26	0.48088	0.49170	0.50232	0.51274	0.52297	0.53301	0.54288	0.55258	0.56210	0.57147
27	0.44651	0.45757	0.46842	0.47907	0.48952	0.49970	0.50986	0.51976	0.52949	0.53905
28	0.41265	0.42397	0.43506	0.44594	0.45662	0.46710	0.47739	0.48750	0.49743	0.50718
29	0.37927	0.39084	0.40218	0.41330	0.42421	0.43491	0.44542	0.45574	0.46587	0.47582
30	0.34630	0.35813	0.36972	0.38108	0.39223	0.40316	0.41389	0.42443	0.43477	0.44493
31	0.31371	0.32580	0.33765	0.34926	0.36065	0.37182	0.38278	0.39353	0.40409	0.41445
32	0.28144	0.29381	0.30592	0.31779	0.32943	0.34084	0.35203	0.36300	0.37378	0.38436
33	0.24947	0.26212	0.27450	0.28664	0.29852	0.31018	0.32161	0.33281	0.34381	0.35461
34	0.21775	0.23069	0.24335	0.25576	0.26790	0.27981	0.29148	0.30292	0.31415	0.32517
35	0.18625	0.19949	0.21244	0.22512	0.23753	0.24970	0.26162	0.27330	0.28476	0.29601
36	0.15493	0.16848	0.18172	0.19469	0.20738	0.21981	0.23199	0.24392	0.25562	0.26710
37	0.12377	0.13763	0.15118	0.16444	0.17741	0.19012	0.20256	0.21475	0.22669	0.23841
38	0.09272	0.10691	0.12078	0.13434	0.14761	0.16059	0.17330	0.18576	0.19796	0.20991
39	0.06177	0.07629	0.09049	0.10436	0.11793	0.13121	0.14420	0.15692	0.16938	0.18159
40	0.03087	0.04575	0.06028	0.07448	0.08836	0.10193	0.11521	0.12821	0.14094	0.15341
41	0.00000	0.01524	0.03013	0.04466	0.05886	0.07275	0.08633	0.09961	0.11262	0.12536
42	---	---	0.00000	0.01488	0.02942	0.04362	0.05751	0.07110	0.08439	0.09740
43	---	---	---	---	0.00000	0.01454	0.02874	0.04263	0.05622	0.06952
44	---	---	---	---	---	---	0.00000	0.01421	0.02810	0.04169
45	---	---	---	---	---	---	---	---	0.00000	0.01389

TABLE A.12.   RANKITS (Normal Order Statistics) (<u>cont.</u>)

Rank	n=91	92	93	94	95	96	97	98	99	100
1	2.47370	2.47764	2.48154	2.48540	2.48920	2.49297	2.49669	2.50036	2.50400	2.50759
2	2.11004	2.11448	2.11887	2.12321	2.12749	2.13172	2.13590	2.14003	2.14411	2.14814
3	1.90536	1.91015	1.91487	1.91953	1.92414	1.92869	1.93318	1.93763	1.94201	1.94635
4	1.75842	1.76348	1.76848	1.77341	1.77828	1.78309	1.78784	1.79254	1.79718	1.80176
5	1.64178	1.64709	1.65232	1.65749	1.66259	1.66763	1.67261	1.67752	1.68238	1.68718
6	1.54396	1.54949	1.55494	1.56033	1.56564	1.57089	1.57607	1.58118	1.58624	1.59123
7	1.45903	1.46476	1.47042	1.47600	1.48151	1.48675	1.49232	1.49762	1.50286	1.50803
8	1.38348	1.38941	1.39526	1.40103	1.40673	1.41235	1.41790	1.42338	1.42879	1.43414
9	1.31511	1.32123	1.32726	1.33321	1.33909	1.34489	1.35061	1.35626	1.36183	1.36734
10	1.25239	1.25869	1.26491	1.27104	1.27708	1.28305	1.28894	1.29475	1.30049	1.30615
11	1.19426	1.20073	1.20712	1.21342	1.21964	1.22577	1.23182	1.23779	1.24368	1.24950
12	1.13990	1.14656	1.15311	1.15958	1.16596	1.17226	1.17847	1.18459	1.19064	1.19661
13	1.08873	1.09555	1.10228	1.10891	1.11546	1.12191	1.12827	1.13455	1.14075	1.14687
14	1.04026	1.04724	1.05415	1.06095	1.06765	1.07426	1.08078	1.08721	1.09356	1.09982
15	0.99413	1.00129	1.00835	1.01531	1.02217	1.02894	1.03561	1.04219	1.04868	1.05509
16	0.95002	0.95735	0.96458	0.97170	0.97872	0.98564	0.99246	0.99919	1.00583	1.01238
17	0.90769	0.91519	0.92258	0.92986	0.93704	0.94411	0.95109	0.95797	0.96475	0.97145
18	0.86693	0.87460	0.88215	0.88959	0.89693	0.90416	0.91129	0.91831	0.92524	0.93208
19	0.82756	0.83540	0.84312	0.85072	0.85822	0.86560	0.87288	0.88006	0.88713	0.89411
20	0.78944	0.79745	0.80533	0.81310	0.82075	0.82829	0.83572	0.84305	0.85027	0.85739
21	0.75243	0.76061	0.76866	0.77659	0.78441	0.79210	0.79968	0.80716	0.81452	0.82179
22	0.71643	0.72478	0.73300	0.74110	0.74907	0.75692	0.76466	0.77228	0.77980	0.78720
23	0.68134	0.68986	0.69825	0.70651	0.71464	0.72266	0.73055	0.73832	0.74598	0.75353
24	0.64706	0.65576	0.66432	0.67275	0.68105	0.68922	0.69727	0.70519	0.71301	0.72070
25	0.61353	0.62241	0.63115	0.63974	0.64821	0.65654	0.66474	0.67282	0.68079	0.68863
26	0.58068	0.58974	0.59865	0.60742	0.61605	0.62454	0.63291	0.64115	0.64926	0.65725
27	0.54845	0.55769	0.56678	0.57572	0.58452	0.59318	0.60170	0.61010	0.61837	0.62651
28	0.51677	0.52620	0.53547	0.54459	0.55356	0.56239	0.57108	0.57963	0.58805	0.59635
29	0.48561	0.49522	0.50468	0.51398	0.52312	0.53212	0.54097	0.54969	0.55827	0.56672
30	0.45491	0.46472	0.47436	0.48384	0.49316	0.50233	0.51136	0.52024	0.52898	0.53758
31	0.42463	0.43464	0.44447	0.45414	0.46364	0.47299	0.48218	0.49123	0.50013	0.50890
32	0.39474	0.40495	0.41498	0.42483	0.43452	0.44404	0.45341	0.46263	0.47170	0.48062
33	0.36520	0.37561	0.38584	0.39588	0.40576	0.41547	0.42501	0.43440	0.44364	0.45273
34	0.33598	0.34660	0.35702	0.36727	0.37733	0.38722	0.39695	0.40652	0.41593	0.42518
35	0.30704	0.31787	0.32850	0.33895	0.34921	0.35929	0.36920	0.37895	0.38853	0.39796
36	0.27835	0.28940	0.30025	0.31090	0.32136	0.33163	0.34173	0.35166	0.36142	0.37102
37	0.24990	0.26117	0.27223	0.28309	0.29375	0.30423	0.31452	0.32464	0.33458	0.34436
38	0.22164	0.23314	0.24443	0.25550	0.26637	0.27705	0.28754	0.29785	0.30797	0.31793
39	0.19356	0.20530	0.21681	0.22810	0.23919	0.25008	0.26077	0.27127	0.28159	0.29173
40	0.16563	0.17761	0.18936	0.20088	0.21219	0.22328	0.23418	0.24488	0.25539	0.26572
41	0.13783	0.15006	0.16205	0.17380	0.18533	0.19665	0.20776	0.21866	0.22937	0.23990
42	0.11014	0.12262	0.13486	0.14685	0.15861	0.17015	0.18148	0.19259	0.20351	0.21423
43	0.08253	0.09528	0.10777	0.12001	0.13201	0.14378	0.15533	0.16666	0.17778	1.18870
44	0.05499	0.06801	0.08076	0.09325	0.10550	0.11750	0.12928	0.14083	0.15217	0.16330
45	0.02748	0.04078	0.05381	0.06656	0.07906	0.09131	0.10332	0.11510	0.12666	0.13800
46	0.00000	0.01359	0.02689	0.03992	0.05267	0.06518	0.07743	0.08944	0.10123	0.11279
47	---	---	0.00000	0.01330	0.02633	0.03909	0.05159	0.06385	0.07586	0.08765
48	---	---	---	---	0.00000	0.01303	0.02579	0.03829	0.05055	0.06257
49	---	---	---	---	---	---	0.00000	0.01276	0.02527	0.03753
50	---	---	---	---	---	---	---	---	0.00000	0.01251

Source: Values extracted by permission from H. L. Harter, *Biometrika* 48 (1961):158-63.

TABLE A.13.   SHAPIRO-WILK TEST FOR NORMALITY

Table A.13.1.   Coefficients of Ordered Differences $(a_{i,n})$

i	n=11	12	13	14	15	16	17	18	19	20
1	.5601	.5475	.5359	.5251	.5150	.5056	.4968	.4886	.4808	.4734
2	.3315	.3325	.3325	.3318	.3306	.3290	.3273	.3253	.3232	.3211
3	.2260	.2347	.2412	.2460	.2495	.2521	.2540	.2553	.2561	.2565
4	.1429	.1586	.1707	.1802	.1878	.1939	.1988	.2027	.2059	.2085
5	.0695	.0922	.1099	.1240	.1353	.1447	.1524	.1587	.1641	.1686
6	...	.0303	.0539	.0727	.0880	.1005	.1109	.1197	.1271	.1334
7	...	...	...	.0240	.0433	.0593	.0725	.0837	.0932	.1013
8	...	...	...	...	...	.0196	.0359	.0496	.0612	.0711
9	...	...	...	...	...	...	...	.0163	.0303	.0422
10	...	...	...	...	...	...	...	...	...	.0140

i	n=21	22	23	24	25	26	27	28	29	30
1	.4643	.4590	.4542	.4493	.4450	.4407	.4366	.4328	.4291	.4254
2	.3185	.3156	.3126	.3098	.3069	.3043	.3018	.2992	.2968	.2944
3	.2578	.2571	.2563	.2554	.2543	.2533	.2522	.2510	.2499	.2487
4	.2119	.2131	.2139	.2145	.2148	.2151	.2152	.2151	.2150	.2148
5	.1736	.1764	.1787	.1807	.1822	.1836	.1848	.1857	.1864	.1870
6	.1399	.1443	.1480	.1512	.1539	.1563	.1584	.1601	.1616	.1630
7	.1092	.1150	.1201	.1245	.1283	.1316	.1346	.1372	.1395	.1415
8	.0804	.0878	.0941	.0997	.1046	.1089	.1128	.1162	.1192	.1219
9	.0530	.0618	.0696	.0764	.0823	.0876	.0923	.0965	.1002	.1036
10	.0263	.0368	.0459	.0539	.0610	.0672	.0728	.0778	.0822	.0862
11	...	.0122	.0228	.0321	.0403	.0476	.0540	.0598	.0650	.0697
12	...	...	...	.0107	.0200	.0284	.0358	.0424	.0483	.0537
13	...	...	...	...	...	.0094	.0178	.0253	.0320	.0381
14	...	...	...	...	...	...	...	.0084	.0159	.0227
15	...	...	...	...	...	...	...	...	...	.0076

Table A.13.1. Coefficients of Ordered Differences $(a_{i,n})$ (<u>cont.</u>)

i	n=31	32	33	34	35	36	37	38	39	40
1	.4220	.4188	.4156	.4127	.4096	.4068	.4040	.4015	.3989	.3964
2	.2921	.2898	.2876	.2854	.2834	.2813	.2794	.2774	.2755	.2737
3	.2475	.2463	.2451	.2439	.2427	.2415	.2403	.2391	.2380	.2368
4	.2145	.2141	.2137	.2132	.2127	.2121	.2116	.2110	.2104	.2098
5	.1874	.1878	.1880	.1882	.1883	.1883	.1883	.1881	.1880	.1878
6	.1641	.1651	.1660	.1667	.1673	.1678	.1683	.1686	.1689	.1691
7	.1433	.1449	.1463	.1475	.1487	.1496	.1505	.1513	.1520	.1526
8	.1243	.1265	.1284	.1301	.1317	.1331	.1344	.1356	.1366	.1376
9	.1066	.1093	.1118	.1140	.1160	.1179	.1196	.1211	.1225	.1237
10	.0899	.0931	.0961	.0988	.1013	.1036	.1056	.1075	.1092	.1108
11	.0739	.0777	.0812	.0844	.0873	.0900	.0924	.0947	.0967	.0986
12	.0585	.0629	.0669	.0706	.0739	.0770	.0798	.0824	.0848	.0870
13	.0435	.0485	.0530	.0572	.0610	.0645	.0677	.0706	.0733	.0759
14	.0298	.0344	.0395	.0441	.0484	.0523	.0559	.0592	.0622	.0651
15	.0144	.0206	.0262	.0314	.0361	.0404	.0444	.0481	.0515	.0546
16	...	.0068	.0131	.0187	.0239	.0287	.0331	.0372	.0409	.0444
17	...	...	...	.0062	.0119	.0172	.0220	.0264	.0305	.0343
18	...	...	...	...	...	.0057	.0110	.0158	.0203	.0244
19	...	...	...	...	...	...	...	.0053	.0101	.0146
20	...	...	...	...	...	...	...	...	...	.0049

Table A.13.1. Coefficients of Ordered Differences $(a_{i,n})$ (cont.)

i	n=41	42	43	44	45	46	47	48	49	50
1	.3940	.3917	.3894	.3872	.3850	.3830	.3808	.3789	.3770	.3751
2	.2719	.2701	.2684	.2667	.2651	.2635	.2620	.2604	.2589	.2574
3	.2357	.2345	.2334	.2323	.2313	.2302	.2291	.2281	.2271	.2260
4	.2091	.2085	.2078	.2072	.2065	.2058	.2052	.2045	.2038	.2032
5	.1876	.1874	.1871	.1868	.1865	.1862	.1859	.1855	.1851	.1847
6	.1693	.1694	.1695	.1695	.1695	.1695	.1695	.1693	.1692	.1691
7	.1531	.1535	.1539	.1542	.1545	.1548	.1550	.1551	.1553	.1554
8	.1384	.1392	.1398	.1405	.1410	.1415	.1420	.1423	.1427	.1430
9	.1249	.1259	.1269	.1278	.1286	.1293	.1300	.1306	.1312	.1317
10	.1123	.1136	.1149	.1160	.1170	.1180	.1189	.1197	.1205	.1212
11	.1004	.1020	.1035	.1049	.1062	.1073	.1085	.1095	.1105	.1113
12	.0891	.0909	.0927	.0943	.0959	.0972	.0986	.0998	.1010	.1020
13	.0782	.0804	.0824	.0842	.0860	.0876	.0892	.0906	.0919	.0932
14	.0677	.0701	.0724	.0745	.0765	.0783	.0801	.0817	.0832	.0846
15	.0575	.0602	.0628	.0651	.0673	.0694	.0713	.0731	.0748	.0764
16	.0476	.0506	.0534	.0560	.0584	.0607	.0628	.0648	.0667	.0685
17	.0379	.0411	.0442	.0471	.0497	.0522	.0546	.0568	.0588	.0608
18	.0283	.0318	.0352	.0383	.0412	.0439	.0465	.0489	.0511	.0532
19	.0188	.0227	.0263	.0296	.0328	.0357	.0385	.0411	.0436	.0459
20	.0094	.0136	.0175	.0211	.0245	.0277	.0307	.0335	.0361	.0386
21	...	.0045	.0087	.0126	.0163	.0197	.0229	.0259	.0288	.0314
22	...	...	...	.0042	.0081	.0118	.0153	.0185	.0215	.0244
23	...	...	...	...	...	.0039	.0076	.0111	.0143	.0174
24	...	...	...	...	...	...	...	.0037	.0071	.0104
25	...	...	...	...	...	...	...	...	...	.0035

Source: Reproduced by permission from S. S. Shapiro and M. B. Wilk, *Biometrika* 52(1965):603-4.

Table A.13.2.   Critical Values for W Statistic*

n	α=0.9	0.5	0.10	0.05	0.02	0.01
11	.973	.940	.876	.850	.817	.792
12	.973	.943	.883	.859	.828	.805
13	.974	.945	.889	.866	.837	.814
14	.975	.947	.895	.874	.846	.825
15	.975	.950	.901	.881	.855	.835
16	.976	.952	.906	.887	.863	.844
17	.977	.954	.910	.892	.869	.851
18	.978	.956	.914	.897	.874	.858
19	.978	.957	.917	.901	.879	.863
20	.979	.959	.920	.905	.884	.868
21	.980	.960	.923	.908	.888	.873
22	.980	.961	.926	.911	.892	.878
23	.981	.962	.928	.914	.895	.881
24	.981	.963	.930	.916	.898	.884
25	.981	.964	.931	.918	.901	.888
26	.982	.965	.933	.920	.904	.891
27	.982	.965	.935	.923	.906	.894
28	.982	.966	.936	.924	.908	.896
29	.982	.966	.937	.926	.910	.898
30	.983	.967	.939	.927	.912	.900
31	.983	.967	.940	.929	.914	.902
32	.983	.968	.941	.930	.915	.904
33	.983	.968	.942	.931	.917	.906
34	.983	.969	.943	.933	.919	.908
35	.984	.969	.944	.934	.920	.910
36	.984	.970	.945	.935	.922	.912
37	.984	.970	.946	.936	.924	.914
38	.984	.971	.947	.938	.925	.916
39	.984	.971	.948	.939	.927	.917
40	.985	.972	.949	.940	.928	.919
41	.985	.972	.950	.941	.929	.920
42	.985	.972	.951	.942	.930	.922
43	.985	.973	.951	.943	.932	.923
44	.985	.973	.952	.944	.933	.924
45	.985	.973	.953	.945	.934	.926
46	.985	.974	.953	.945	.935	.927
47	.985	.974	.954	.946	.936	.928
48	.985	.974	.954	.947	.937	.929
49	.985	.974	.955	.947	.937	.929
50	.985	.974	.955	.947	.938	.930

Source: Reproduced by permission from S. S. Shapiro and M. B. Wilk, *Biometrika* 52(1965):605.
   *Nonnormality is indicated when the W statistic is *smaller* than the appropriate critical value.

TABLE A.14.   EXPECTED RATIO OF RANGE TO STANDARD DEVIATION IN NORMAL SAMPLES

n	0	1	2	3	4	5	6	7	8	9
0	...	...	1.128	1.693	2.059	2.326	2.534	2.704	2.847	2.970
10	3.078	3.173	3.258	3.336	3.407	3.472	3.532	3.588	3.640	3.689
20	3.735	3.778	3.819	3.858	3.895	3.931	3.964	3.997	4.027	4.057
30	4.086	4.113	4.139	4.165	4.189	4.213	4.236	4.259	4.280	4.301
40	4.322	4.341	4.361	4.379	4.398	4.415	4.433	4.450	4.466	4.482
50	4.498	4.514	4.529	4.543	4.558	4.572	4.586	4.599	4.613	4.626
60	4.639	4.651	4.663	4.676	4.687	4.699	4.711	4.722	4.733	4.744
70	4.755	4.765	4.776	4.786	4.796	4.806	4.816	4.825	4.835	4.844
80	4.854	4.863	4.872	4.881	4.889	4.898	4.906	4.915	4.923	4.931
90	4.939	4.947	4.955	4.963	4.971	4.978	4.986	4.993	5.001	5.008
100	5.015	5.022	5.029	5.036	5.043	5.050	5.057	5.063	5.070	5.076
110	5.083	5.089	5.096	5.102	5.108	5.114	5.120	5.126	5.132	5.138
120	5.144	5.150	5.156	5.161	5.167	5.173	5.178	5.184	5.189	5.195
130	5.200	5.205	5.211	5.216	5.221	5.226	5.231	5.236	5.241	5.246
140	5.251	5.256	5.261	5.266	5.271	5.275	5.280	5.285	5.289	5.294
150	5.298	5.303	5.308	5.312	5.316	5.321	5.325	5.330	5.334	5.338
160	5.342	5.347	5.351	5.355	5.359	5.363	5.367	5.371	5.375	5.379
170	5.383	5.387	5.391	5.395	5.399	5.403	5.407	5.411	5.414	5.418
180	5.422	5.426	5.429	5.433	5.437	5.440	5.444	5.447	5.451	5.454
190	5.458	5.461	5.465	5.468	5.472	5.475	5.479	5.482	5.485	5.489

n	0	10	20	30	40	50	60	70	80	90
200	5.492	5.524	5.555	5.584	5.612	5.638	5.664	5.688	5.711	5.734
300	5.756	5.776	5.797	5.816	5.835	5.853	5.871	5.888	5.904	5.921
400	5.936	5.952	5.967	5.981	5.995	6.009	6.023	6.036	6.049	6.061
500	6.073	6.085	6.097	6.109	6.120	6.131	6.142	6.153	6.163	6.173
600	6.183	6.193	6.203	6.213	6.222	6.231	6.240	6.249	6.258	6.267
700	6.275	6.283	6.292	6.300	6.308	6.316	6.324	6.331	6.339	6.346
800	6.354	6.361	6.368	6.375	6.382	6.389	6.396	6.402	6.409	6.416
900	6.422	6.429	6.435	6.441	6.447	6.453	6.459	6.465	6.471	6.477

Source: Values extracted by permission from E. S. Pearson and H. O. Hartley, eds., *Biometrika tables for statisticians*, vol. 1, 2d ed. (Cambridge Univ. Press), p. 174.   The ratios may be used to estimate the standard deviation of a normal population, given sample or subjective information about the range.

$$SD = \frac{range}{\#}$$

A.15. POWER FUNCTION AND SAMPLE SIZES FOR TESTS OF HYPOTHESES

In each of Figs. A.15.1-A.15.21, the abscissa scale is a function of the difference one wishes to detect (as explained in the text), and the ordinate scale is the probability of a Type II error (1 Power). The numbered curve most nearly coinciding with the intersection of the relative difference to be detected and the allowable magnitude of Type II error represents the approximate sample size required. In some cases it is the total size of sample, in others the replication per treatment group or number of degrees of freedom for error. See appropriate sections of the text.

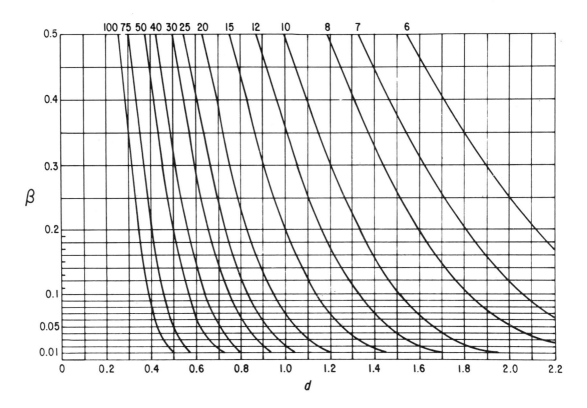

Fig. A.15.1. Sample size for two-sided t test (α = 0.01). Drawn from values from tables by J. L. Hodges, Jr., and E. L. Lehmann, *Ann. Math. Stat.* 39(1968):1629-37.

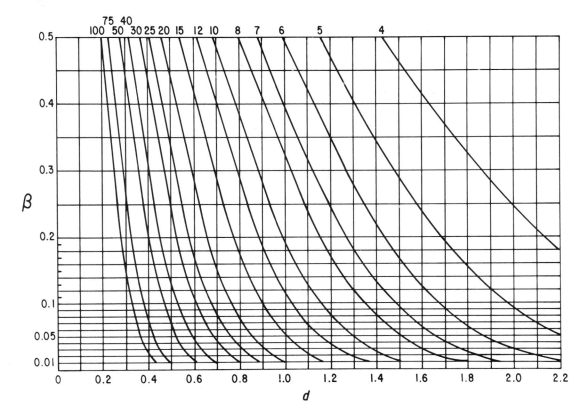

Fig. A.15.2.   Sample size for two-sided t test (α = 0.05).   Drawn from tables
by Hodges and Lehmann, *Ann. Math. Stat.* 39:1629-37.

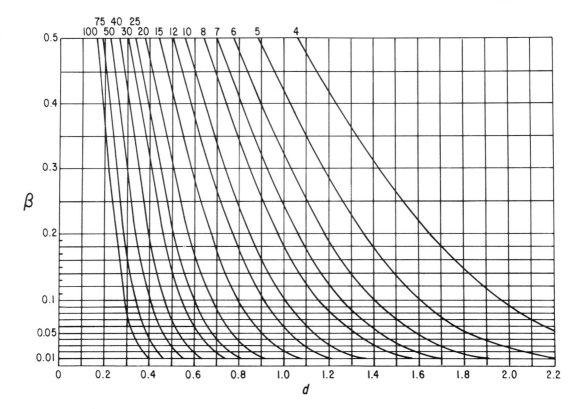

Fig. A.15.3. Sample size for two-sided *t* test (α = 0.10). Drawn from tables
by Hodges and Lehmann, *Ann. Math. Stat.* 39:1629-37.

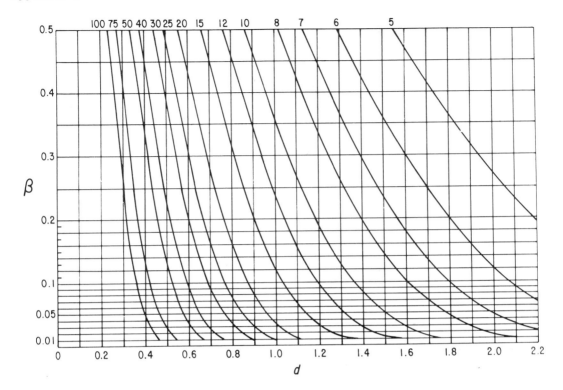

Fig. A.15.4. Sample size for one-sided *t* test (α = 0.01). Drawn from tables by Hodges and Lehmann, *Ann. Math. Stat.* 39:1629-37.

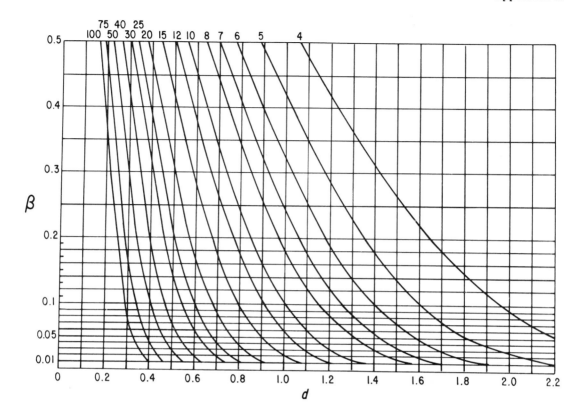

Fig. A.15.5.   Sample size for one-sided t test (α = 0.05).   Drawn from tables by Hodges and Lehmann, *Ann. Math. Stat.* 39:1629-37.

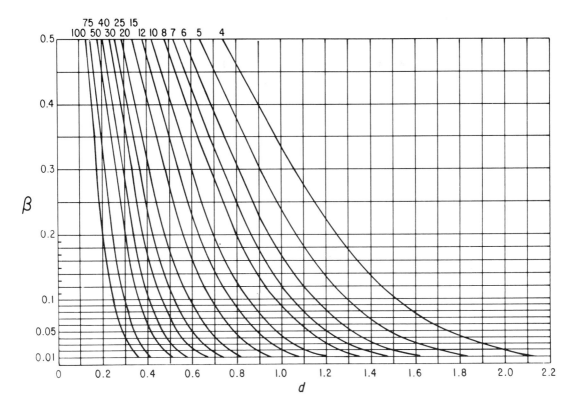

Fig. A.15.6. Sample size for one-sided t test ($\alpha = 0.10$). Drawn from tables by Hodges and Lehmann, *Ann. Math. Stat.* 39:1629-37.

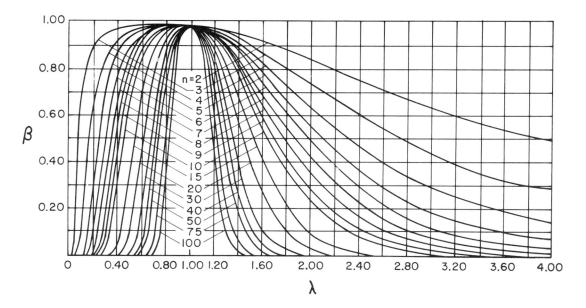

Fig. A.15.7. Sample size for two-sided chi-square test (α = 0.01), from Al-
bert H. Bowker and Gerald J. Lieberman, 1972, *Engineering statistics*, 2d ed., p. 209. Reprinted by permission of the authors
and Prentice-Hall, Englewood Cliffs, N.J.

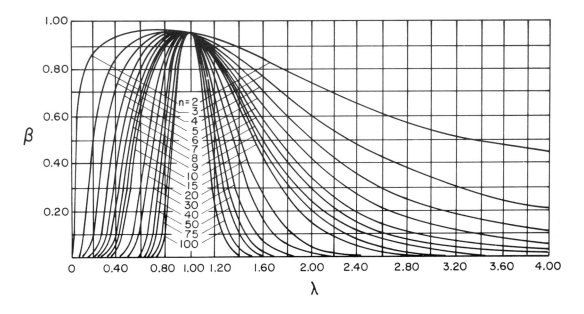

Fig. A.15.8. Sample size for two-sided chi-square test (α = 0.05). Reprinted
by permission from Bowker and Lieberman, *Engineering statistics*,
p. 209.

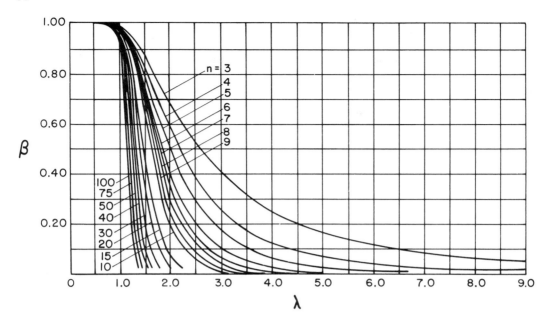

Fig. A.15.9. Sample size for one-sided (upper-tail) chi-square test (α = 0.01). Reprinted by permission from Bowker and Lieberman, *Engineering statistics*, p. 211.

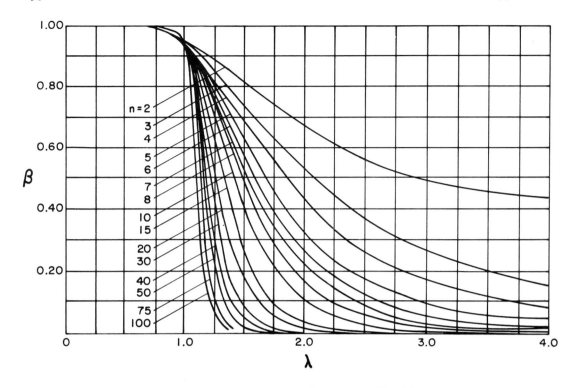

Fig. A.15.10.   Sample size for one-sided (upper-tail) chi-square test
                ($\alpha = 0.05$). Reproduced by permission from C. L. Ferris, F. E.
                Grubbs, and C. L. Weaver, *Ann. Math. Stat.* 17(1946):181.

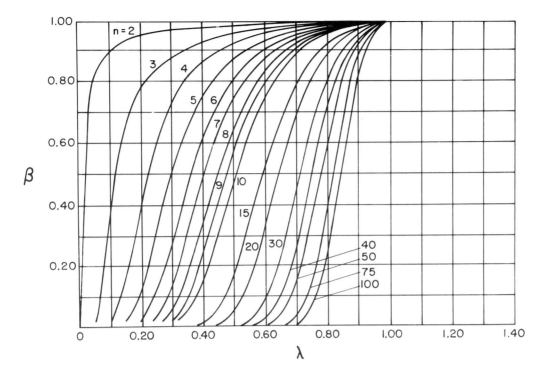

Fig. A.15.11.   Sample size for one-sided (lower-tail) chi-square test
(α = 0.01).   Reprinted by permission from Bowker and Lieberman,
Engineering statistics, p. 213.

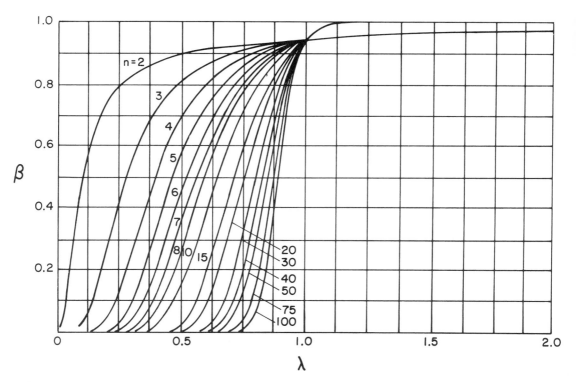

Fig. A.15.12. Sample size for one-sided (lower-tail) chi-square test
(α = 0.05). Reproduced by permission from Ferris, Grubbs, and
Weaver, *Ann. Math. Stat.* 17:183.

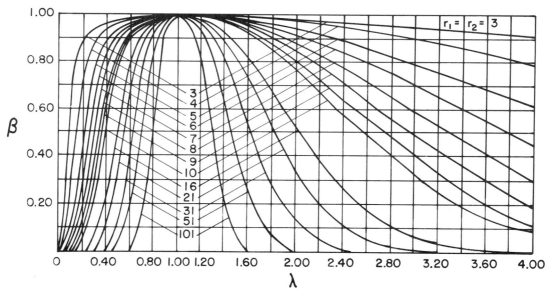

Fig. A.15.13. Sample size for two-sided *F* test (α = 0.01). Reprinted by permission from Bowker and Lieberman, *Engineering statistics*, p.
256.

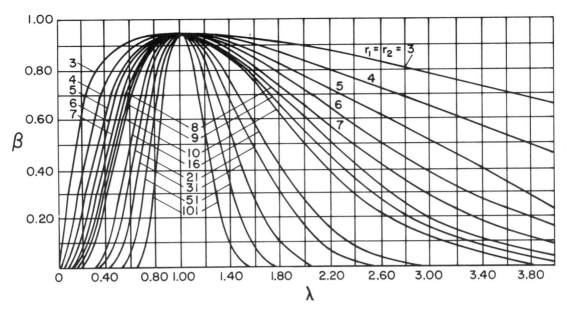

Fig. A.15.14. Sample size for two-sided F test ($\alpha = 0.05$). Reprinted by permission from Bowker and Lieberman, *Engineering statistics*, p. 257.

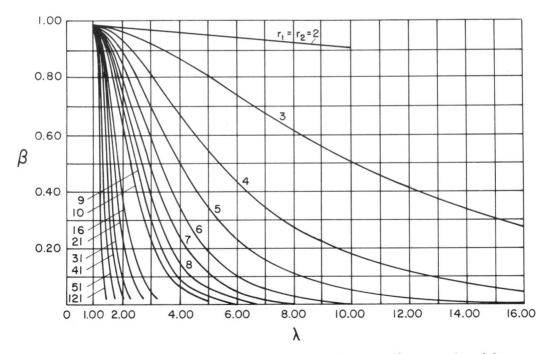

Fig. A.15.15. Sample size for one-sided F test ($\alpha = 0.01$). Reprinted by permission from Bowker and Lieberman, *Engineering statistics*, p. 259.

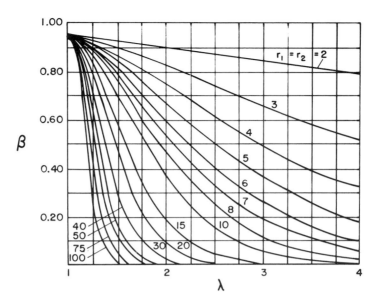

Fig. A.15.16.   Sample size for one-sided *F* test (α = 0.05).   Reproduced by
permission from Ferris, Grubbs, and Weaver, *Ann. Math. Stat.*
17:185.

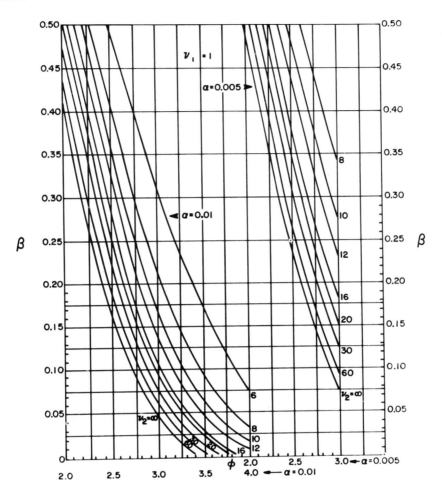

Fig. A.15.17. Sample size (ν_2 df for error) for power of f test; analysis of variance of fixed effects ($\nu_1 = 1$, $\alpha = 0.005$ or 0.01). Drawn using values from tables by M. L. Tiku, *J. Am. Stat. Assoc.* 62(1967):527–32.

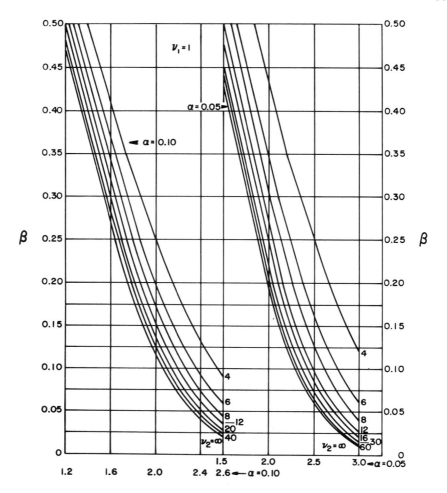

Fig. A.15.18. Sample size (ν_2 df for error) for power of f test; analysis of
variance of fixed effects ($\nu_1 = 1$, $\alpha = 0.05$ or 0.10). Drawn
from tables by Tiku, *J. Am. Stat. Assoc.* 62:535–37.

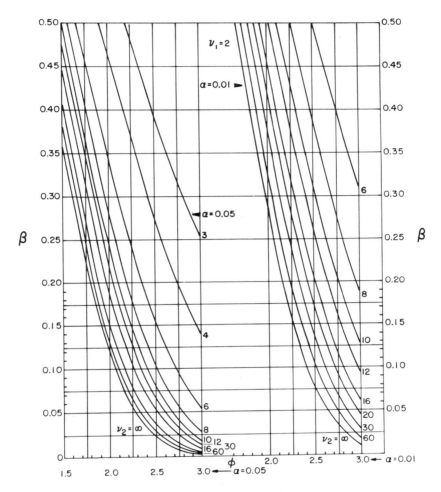

Fig. A.15.19. Sample size (ν_2 df for error) for power of f test; analysis of
variance of fixed effects ($\nu_1 = 2$, $\alpha = 0.01$ or 0.05). Drawn
from tables by Tiku, *J. Am. Stat. Assoc.* 62:529–37.

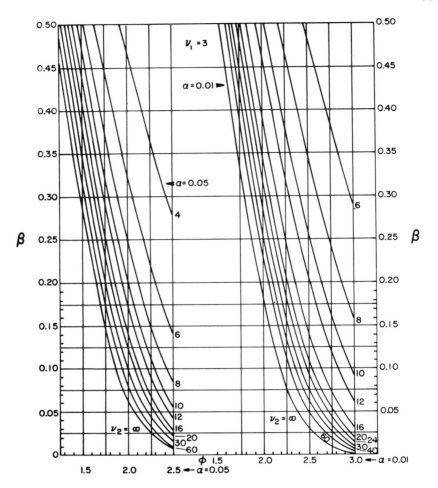

Fig. A.15.20. Sample size (ν_2 df for error) for power of f test; analysis of variance of fixed effects ($\nu_1 = 3$, $\alpha = 0.01$ or 0.05). Drawn from tables by Tiku, *J. Am. Stat. Assoc.* 62:529-37.

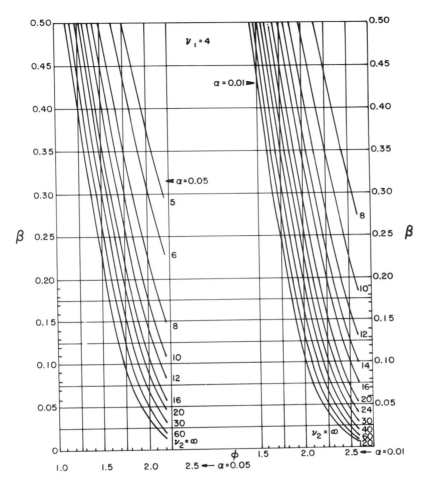

Fig. A.15.21.   Sample size (ν_2 df for error) for power of f test; analysis of variance of fixed effects ($\nu_1 = 4$, α = 0.01 or 0.05).   Drawn from tables by Tiku, *J. Am. Stat. Assoc.* 62:529-37.

TABLE A.16. CRITICAL VALUES FOR TESTING ONE OUTLYING OBSERVATION

n	α=0.01	0.05	0.10	n	α=0.01	0.05	0.10
...	...	...	...	32	3.135	2.773	2.591
...	...	...	...	34	3.164	2.799	2.616
3	1.155	1.153	1.148	36	3.191	2.823	2.639
4	1.492	1.463	1.425	38	3.216	2.846	2.661
5	1.749	1.672	1.602	40	3.240	2.866	2.682
6	1.944	1.822	1.729	42	3.261	2.887	2.700
7	2.097	1.938	1.828	44	3.282	2.905	2.719
8	2.221	2.032	1.909	46	3.302	2.923	2.736
9	2.323	2.110	1.977	48	3.319	2.940	2.753
10	2.410	2.176	2.036	50	3.336	2.956	2.768
11	2.485	2.234	2.088	55	3.376	2.992	2.804
12	2.550	2.285	2.134	60	3.411	3.025	2.837
13	2.607	2.331	2.175	65	3.442	3.055	2.866
14	2.659	2.371	2.213	70	3.471	3.082	2.893
15	2.705	2.409	2.247	75	3.496	3.107	2.917
16	2.747	2.443	2.279	80	3.521	3.130	2.940
17	2.785	2.475	2.309	85	3.543	3.151	2.961
18	2.821	2.504	2.335	90	3.563	3.171	2.981
19	2.854	2.532	2.361	95	3.582	3.189	3.000
20	2.884	2.557	2.385	100	3.600	3.207	3.017
21	2.912	2.580	2.408	105	3.617	3.224	3.033
22	2.939	2.603	2.429	110	3.632	3.239	3.049
23	2.963	2.624	2.448	115	3.647	3.254	3.064
24	2.987	2.644	2.467	120	3.662	3.267	3.078
25	3.009	2.663	2.486	125	3.675	3.281	3.092
26	3.029	2.681	2.502	130	3.688	3.294	3.104
27	3.049	2.698	2.519	135	3.700	3.306	3.116
28	3.068	2.714	2.534	140	3.712	3.318	3.129
29	3.085	2.730	2.549	145	3.723	3.328	3.140
30	3.103	2.745	2.563				

Source: Values extracted by permission from F. E. Grubbs and G. Beck, *Technometrics* 14(1972):848-50.

TABLE A.17.   CRITICAL VALUES* FOR TESTING TWO LARGEST OR TWO SMALLEST OBSERVATIONS

n	α=0.01	0.05	0.10	n	α=0.01	0.05	0.10
...	...	...	...	32	.546	.618	.653
...	...	...	...	34	.565	.633	.667
...	...	...	...	36	.581	.647	.680
4	.000	.001	.003	38	.596	.660	.692
5	.004	.018	.038	40	.610	.672	.702
6	.019	.056	.092	42	.623	.683	.712
7	.044	.102	.148	44	.636	.694	.722
8	.075	.148	.199	46	.647	.703	.730
9	.108	.191	.245	48	.657	.712	.738
10	.141	.230	.286	50	.667	.720	.746
11	.174	.267	.323	55	.689	.739	.763
12	.204	.300	.355	60	.709	.755	.777
13	.233	.330	.384	65	.725	.769	.790
14	.260	.357	.411	70	.740	.781	.801
15	.286	.382	.434	75	.753	.792	.811
16	.310	.405	.456	80	.765	.802	.820
17	.332	.426	.476	85	.776	.811	.828
18	.353	.446	.494	90	.785	.819	.835
19	.372	.464	.511	95	.794	.826	.841
20	.391	.480	.527	100	.802	.833	.848
21	.408	.496	.542	105	.809	.839	.853
22	.424	.511	.555	110	.816	.845	.858
23	.440	.524	.568	115	.822	.850	.863
24	.454	.537	.580	120	.828	.855	.867
25	.468	.550	.591	125	.834	.859	.871
26	.481	.561	.601	130	.839	.864	.875
27	.493	.572	.611	135	.844	.868	.879
28	.505	.582	.620	140	.848	.871	.882
29	.516	.592	.629	145	.852	.875	.885
30	.527	.601	.638				

*Note that a value of the test statistic *smaller* than the critical value is significant.

Source: Values extracted by permission from Grubbs and Beck, *Technometrics* 14(1972):851-53.

TABLE A.18.   UPPER PERCENTAGE POINTS OF HOTELLING'S T^2 DISTRIBUTION (1–CDF)*

ν	α	p=1	2	3	4	5	6	7	8	9
2	.01	98.503	---	---	---	---	---	---	---	---
	.05	18.513	---	---	---	---	---	---	---	---
	.10	8.526	---	---	---	---	---	---	---	---
3	.01	34.116	297.00	---	---	---	---	---	---	---
	.05	10.128	57.000	---	---	---	---	---	---	---
	.10	5.538	27.000	---	---	---	---	---	---	---
4	.01	21.198	82.177	595.00	---	---	---	---	---	---
	.05	7.709	25.472	114.99	---	---	---	---	---	---
	.10	4.545	14.566	54.971	---	---	---	---	---	---
5	.01	16.258	45.000	147.28	992.49	---	---	---	---	---
	.05	6.608	17.361	46.383	192.47	---	---	---	---	---
	.10	4.060	10.811	26.954	92.434	---	---	---	---	---
6	.01	13.745	31.857	75.125	229.68	1489.5	---	---	---	---
	.05	5.987	13.887	29.661	72.937	289.45	---	---	---	---
	.10	3.776	9.071	18.859	42.741	139.39	---	---	---	---
7	.01	12.246	25.491	50.652	111.84	329.43	2086.0	---	---	---
	.05	5.591	12.001	22.720	44.718	105.16	405.92	---	---	---
	.10	3.589	8.081	15.202	28.751	61.940	195.84	---	---	---
8	.01	11.259	21.821	39.118	72.908	155.22	446.57	2782.0	---	---
	.05	5.318	10.828	19.028	33.230	62.561	143.05	541.89	---	---
	.10	3.458	7.446	13.155	22.529	40.506	84.556	261.77	---	---
9	.01	10.561	19.460	32.598	54.890	98.703	205.29	581.11	3577.5	---
	.05	5.117	10.033	16.766	27.202	45.453	83.202	186.62	697.36	---
	.10	3.360	7.005	11.857	19.085	31.077	54.132	110.59	337.20	---
10	.01	10.044	17.826	28.466	44.838	72.882	128.07	262.08	733.04	4472.5
	.05	4.965	9.459	15.248	23.545	36.561	59.403	106.65	235.87	872.32
	.10	3.285	6.681	10.964	16.917	25.896	40.854	69.63	140.04	422.12

*Values in this table may be used as critical values in the multivariate test of a p–dimensional vector of means if $\nu > p$ df.

TABLE A.18.   UPPER PERCENTAGE POINTS OF HOTELLING'S T^2 DISTRIBUTION (1–CDF)
              (cont.)

ν	α	p=1	2	3	4	5	6	7	8	9
11	.01	9.646	16.631	25.637	38.533	58.618	93.127	161.02	325.58	902.39
	.05	4.844	9.026	14.163	21.108	31.205	47.123	75.088	132.90	290.81
	.10	3.225	6.434	10.314	15.435	22.655	33.600	51.866	87.009	172.92
12	.01	9.330	15.722	23.588	34.251	49.739	73.969	115.64	197.56	395.80
	.05	4.747	8.689	13.350	19.376	27.656	39.764	58.893	92.512	161.97
	.10	3.177	6.239	9.820	14.361	20.448	29.082	42.202	64.114	106.26
13	.01	9.074	15.008	22.041	31.171	43.745	62.114	90.907	140.43	237.69
	.05	4.667	8.418	12.719	18.086	25.145	34.911	49.232	71.878	111.68
	.10	3.136	6.081	9.432	13.548	18.854	26.016	36.204	51.706	77.601
14	.01	8.862	14.433	20.834	28.857	39.454	54.150	75.676	109.44	167.50
	.05	4.600	8.197	12.216	17.089	23.281	31.488	42.881	59.612	86.079
	.10	3.102	5.951	9.119	12.912	17.651	23.808	32.146	44.025	62.113
15	.01	8.683	13.960	19.867	27.060	36.246	48.472	65.483	90.433	129.58
	.05	4.543	8.012	11.806	16.296	21.845	28.955	38.415	51.572	70.907
	.10	3.073	5.842	8.862	12.401	16.713	22.145	29.229	38.840	52.547
16	.01	8.531	13.566	19.076	25.626	33.762	44.240	58.241	77.755	106.39
	.05	4.494	7.856	11.465	15.651	20.706	27.008	35.117	45.932	60.986
	.10	3.048	5.750	8.648	11.981	15.960	20.850	27.036	35.120	46.102
17	.01	8.400	13.231	18.418	24.458	31.788	40.975	52.858	68.771	90.969
	.05	4.451	7.722	11.177	15.117	19.782	25.467	32.588	41.775	54.041
	.10	3.026	5.670	8.465	11.631	15.344	19.814	25.331	32.329	41.486
18	.01	8.285	12.943	17.861	23.487	30.182	38.385	48.715	62.109	80.067
	.05	4.414	7.606	10.931	14.667	19.017	24.219	30.590	38.592	48.930
	.10	3.007	5.600	8.309	11.335	14.830	18.966	23.969	30.161	38.026
19	.01	8.185	12.694	17.385	22.670	28.852	36.283	45.435	56.992	71.999
	.05	4.381	7.504	10.719	14.283	18.375	23.189	28.975	36.082	45.023
	.10	2.990	5.539	8.173	11.081	14.396	18.261	22.857	28.431	35.343
20	.01	8.096	12.476	16.973	21.972	27.734	34.546	42.779	52.948	65.813
	.05	4.351	7.415	10.533	13.952	17.828	22.324	27.642	34.054	41.946
	.10	2.975	5.485	8.053	10.860	14.023	17.665	21.931	27.020	33.203

TABLE A.18.   UPPER PERCENTAGE POINTS OF HOTELLING'S T^2 DISTRIBUTION (1-CDF)
(cont.)

ν	α	p=1	2	3	4	5	6	7	8	9
21	.01	8.017	12.283	16.613	21.369	26.781	33.088	40.587	49.679	60.932
	.05	4.325	7.335	10.370	13.663	17.356	21.588	26.525	32.384	39.463
	.10	2.961	5.437	7.948	10.667	13.701	17.154	21.150	25.847	31.459
22	.01	7.945	12.111	16.296	20.843	25.959	31.847	38.750	46.986	56.991
	.05	4.301	7.264	10.225	13.409	16.945	20.954	25.576	30.985	37.419
	.10	2.949	5.394	7.854	10.497	13.419	16.713	20.482	24.857	30.011
23	.01	7.881	11.958	16.015	20.381	25.244	30.779	37.188	44.730	53.748
	.05	4.279	7.200	10.095	13.184	16.585	20.403	24.759	29.798	35.709
	.10	2.937	5.355	7.770	10.345	13.170	16.327	19.904	24.012	28.790
24	.01	7.823	11.820	15.763	19.972	24.616	29.850	35.846	42.816	51.036
	.05	4.260	7.142	9.979	12.983	16.265	19.920	24.049	28.777	34.258
	.10	2.927	5.320	7.695	10.210	12.949	15.987	19.400	23.281	27.747
25	.01	7.770	11.695	15.538	19.606	24.060	29.036	34.680	41.171	48.736
	.05	4.242	7.089	9.874	12.803	15.981	19.492	23.427	27.891	33.013
	.10	2.918	5.288	7.626	10.088	12.752	15.685	18.955	22.643	26.846
26	.01	7.721	11.581	15.334	19.279	23.565	28.316	33.659	39.745	46.762
	.05	4.225	7.041	9.779	12.641	15.726	19.112	22.878	27.114	31.932
	.10	2.909	5.259	7.564	9.977	12.574	15.415	18.561	22.082	26.061
27	.01	7.677	11.478	15.149	18.983	23.121	27.675	32.756	38.496	45.051
	.05	4.210	6.997	9.692	12.493	15.496	18.770	22.388	26.428	30.985
	.10	2.901	5.232	7.507	9.877	12.414	15.173	18.209	21.584	25.370
28	.01	7.636	11.383	14.980	18.715	22.721	27.101	31.954	37.393	43.554
	.05	4.196	6.957	9.612	12.359	15.287	18.463	21.950	25.818	30.149
	.10	2.894	5.207	7.455	9.785	12.268	14.954	17.893	21.140	24.757
29	.01	7.598	11.295	14.825	18.471	22.359	26.584	31.236	36.414	42.234
	.05	4.183	6.919	9.539	12.236	15.097	18.184	21.555	25.272	29.407
	.10	2.887	5.184	7.407	9.701	12.135	14.755	17.607	20.741	24.211
30	.01	7.562	11.215	14.683	18.247	22.029	26.116	30.589	35.538	41.062
	.05	4.171	6.885	9.471	12.123	14.924	17.931	21.198	24.781	28.742
	.10	2.881	5.163	7.363	9.624	12.013	14.573	17.348	20.380	23.720

TABLE A.18.   UPPER PERCENTAGE POINTS OF HOTELLING'S T^2 DISTRIBUTION (1-CDF)
              (cont.)

ν	α	p=1	2	3	4	5	6	7	8	9
35	.01	7.419	10.890	14.117	17.366	20.743	24.314	28.135	32.259	36.743
	.05	4.121	6.744	9.200	11.674	14.240	16.944	19.823	22.913	26.252
	.10	2.855	5.077	7.184	9.316	11.530	13.862	16.343	19.001	21.866
40	.01	7.314	10.655	13.715	16.750	19.858	23.094	26.502	30.120	33.984
	.05	4.085	6.642	9.005	11.356	13.762	16.264	18.890	21.668	24.624
	.10	2.835	5.013	7.054	9.095	11.190	13.369	15.657	18.074	20.642
45	.01	7.234	10.478	13.414	16.295	19.211	22.214	25.340	28.617	32.073
	.05	4.057	6.564	8.859	11.118	13.409	15.767	18.217	20.781	23.477
	.10	2.820	4.965	6.957	8.930	10.937	13.006	15.158	17.408	19.773
50	.01	7.171	10.340	13.181	15.945	18.718	21.550	24.470	27.504	30.673
	.05	4.034	6.503	8.744	10.934	13.138	15.388	17.709	20.117	22.627
	.10	2.809	4.927	6.880	8.802	10.743	12.729	14.779	16.907	19.125
55	.01	7.119	10.228	12.995	15.667	18.331	21.030	23.795	26.647	29.603
	.05	4.016	6.454	8.652	10.787	12.923	15.090	17.311	19.600	21.972
	.10	2.799	4.896	6.818	8.699	10.588	12.510	14.481	16.516	18.623
60	.01	7.077	10.137	12.843	15.442	18.018	20.613	23.257	25.967	28.760
	.05	4.001	6.413	8.577	10.668	12.748	14.850	16.992	19.188	21.451
	.10	2.791	4.871	6.768	8.616	10.462	12.332	14.242	16.202	18.223
70	.01	7.011	9.996	12.611	15.098	17.543	19.986	22.451	24.957	27.515
	.05	3.978	6.350	8.460	10.484	12.482	14.485	16.510	18.571	20.676
	.10	2.779	4.831	6.690	8.487	10.270	12.062	13.880	15.731	17.625
80	.01	6.963	9.892	12.440	14.849	17.201	19.536	21.877	24.242	26.642
	.05	3.960	6.303	8.375	10.350	12.289	14.222	16.165	18.130	20.127
	.10	2.769	4.802	6.632	8.392	10.129	11.867	13.619	15.394	17.200
90	.01	6.925	9.813	12.310	14.660	16.942	19.197	21.448	23.710	25.995
	.05	3.947	6.267	8.309	10.248	12.142	14.022	15.905	17.801	19.718
	.10	2.762	4.780	6.588	8.320	10.023	11.719	13.422	15.141	16.882

TABLE A.18.   UPPER PERCENTAGE POINTS OF HOTELLING'S T^2 DISTRIBUTION (1-CDF)
(cont.)

ν	α	p=1	2	3	4	5	6	7	8	9
100	.01	6.895	9.750	12.208	14.511	16.740	18.934	21.115	23.299	25.496
	.05	3.936	6.239	8.257	10.167	12.027	13.867	15.702	17.544	19.401
	.10	2.756	4.762	6.553	8.263	9.939	11.603	13.268	14.944	16.635
110	.01	6.871	9.699	12.125	14.391	16.577	18.722	20.849	22.972	25.101
	.05	3.927	6.216	8.215	10.102	11.934	13.741	15.540	17.340	19.149
	.10	2.752	4.747	6.524	8.217	9.871	11.509	13.145	14.786	16.438
120	.01	6.851	9.657	12.057	14.292	16.444	18.549	20.632	22.705	24.779
	.05	3.920	6.196	8.181	10.048	11.858	13.639	15.407	17.172	18.943
	.10	2.748	4.735	6.501	8.179	9.815	11.432	13.043	14.657	16.278
150	.01	6.807	9.565	11.909	14.079	16.156	18.178	20.167	22.137	24.096
	.05	3.904	6.155	8.105	9.931	11.693	13.417	15.121	16.814	18.504
	.10	2.739	4.708	6.449	8.096	9.694	11.266	12.826	14.380	15.934
200	.01	6.763	9.474	11.764	13.871	15.877	17.819	19.720	21.592	23.446
	.05	3.888	6.113	8.031	9.817	11.531	13.202	14.845	16.469	18.083
	.10	2.731	4.682	6.399	8.015	9.576	11.105	12.614	14.112	15.603
400	.01	6.699	9.341	11.551	13.569	15.473	17.303	19.080	20.818	22.525
	.05	3.865	6.052	7.922	9.650	11.297	12.890	14.447	15.975	17.484
	.10	2.718	4.643	6.324	7.895	9.403	10.870	12.309	13.727	15.131
1000	.01	6.660	9.262	11.426	13.392	15.239	17.006	18.713	20.376	22.003
	.05	3.851	6.015	7.857	9.552	11.160	12.710	14.217	15.692	17.141
	.10	2.711	4.620	6.280	7.825	9.303	10.734	12.132	13.506	14.859
∞	.01	6.635	9.210	11.345	13.277	15.086	16.812	18.475	20.090	21.666
	.05	3.841	5.991	7.815	9.488	11.070	12.592	14.067	15.507	16.919
	.10	2.706	4.605	6.251	7.779	9.236	10.645	12.017	13.362	14.684

TABLE A.18. UPPER PERCENTAGE POINTS OF HOTELLING'S T^2 DISTRIBUTION (1-CDF)
(cont.)

ν	α	p=10	11	12	13	14	15	16	17	18
11	.01	5467.0	---	---	---	---	---	---	---	---
	.05	1066.8	---	---	---	---	---	---	---	---
	.10	516.54	---	---	---	---	---	---	---	---
12	.01	1089.1	6560.9	---	---	---	---	---	---	---
	.05	351.42	1280.7	---	---	---	---	---	---	---
	.10	209.22	620.44	---	---	---	---	---	---	---
13	.01	472.74	1293.3	7754.4	---	---	---	---	---	---
	.05	193.84	417.72	1514.2	---	---	---	---	---	---
	.10	127.40	248.94	733.83	---	---	---	---	---	---
14	.01	281.43	556.41	1514.9	9047.4	---	---	---	---	---
	.05	132.58	228.53	489.70	1767.1	---	---	---	---	---
	.10	92.327	150.41	292.08	856.72	---	---	---	---	---
15	.01	196.85	328.77	646.81	1753.9	10440.	---	---	---	---
	.05	101.50	155.23	266.03	567.36	2039.6	---	---	---	---
	.10	73.423	108.29	175.30	338.64	989.10	---	---	---	---
16	.01	151.32	228.49	379.71	743.94	2010.3	11932.	---	---	---
	.05	83.121	118.14	179.62	306.34	650.71	2331.5	---	---	---
	.10	61.772	85.639	125.50	202.07	388.62	1131.0	---	---	---
17	.01	123.55	174.66	262.42	434.26	847.79	2284.1	13523.	---	---
	.05	71.127	96.253	136.00	205.76	349.46	739.74	2642.9	---	---
	.10	53.933	71.699	98.760	143.95	230.72	442.03	1282.3	---	---
18	.01	105.13	141.92	199.62	298.64	492.41	958.38	2575.4	15214.	---
	.05	62.746	81.996	110.30	155.08	233.64	395.40	834.46	2973.9	---
	.10	48.326	62.334	82.330	112.79	163.64	261.25	498.86	1443.2	---
19	.01	92.134	120.24	161.50	226.18	337.15	554.17	1075.7	2884.0	17005.
	.05	56.587	72.047	93.592	125.28	175.38	263.27	444.15	934.86	3324.3
	.10	44.129	55.643	71.306	93.666	127.72	184.57	293.66	559.10	1613.5
20	.01	82.532	104.97	136.30	182.29	254.36	377.95	619.53	1199.7	3210.1
	.05	51.884	64.745	81.945	105.92	141.17	196.90	294.64	495.72	1040.9
	.10	40.877	50.640	63.437	80.849	105.71	143.56	206.74	327.95	622.78

TABLE A.18.   UPPER PERCENTAGE POINTS OF HOTELLING'S T^2 DISTRIBUTION (1-CDF)
(cont.)

ν	α	p=10	11	12	13	14	15	16	17	18
21	.01	75.181	93.711	118.59	153.32	204.29	284.14	421.04	688.50	1330.5
	.05	48.184	59.177	73.407	92.442	118.97	157.98	219.65	327.76	550.10
	.10	38.286	46.766	57.558	71.708	90.963	118.45	160.31	230.15	364.12
22	.01	69.389	85.100	105.54	132.98	171.29	227.50	315.54	446.42	761.08
	.05	45.202	54.800	66.902	82.573	103.54	132.76	175.72	243.62	362.62
	.10	36.175	43.682	53.009	64.885	80.458	101.65	131.90	177.96	254.80
23	.01	64.719	78.323	95.571	118.01	148.15	190.21	251.92	348.55	514.10
	.05	42.750	51.274	61.793	75.060	92.244	115.23	147.28	194.38	268.80
	.10	34.424	41.171	49.391	59.609	72.621	89.686	112.91	146.06	196.52
24	.01	60.879	72.865	87.736	106.60	131.14	164.09	210.09	277.56	383.18
	.05	40.699	48.378	57.681	69.165	83.653	102.42	127.53	162.52	213.96
	.10	32.949	39.090	46.447	55.415	66.565	80.767	99.392	124.74	160.92
25	.01	57.671	68.382	81.432	97.630	118.18	144.92	180.82	230.93	304.40
	.05	38.961	45.958	54.305	64.423	76.916	92.681	113.10	140.42	178.50
	.10	31.690	37.337	44.008	52.005	61.754	73.878	89.322	109.58	137.14
26	.01	54.953	64.639	76.258	90.421	108.00	130.31	159.34	198.32	252.72
	.05	37.469	43.908	51.487	60.533	71.501	85.048	102.14	124.29	153.92
	.10	30.603	35.842	41.955	49.180	57.843	68.408	81.548	98.287	120.24
27	.01	52.622	61.470	71.942	84.509	99.834	118.86	143.00	174.42	216.61
	.05	36.176	42.149	49.099	57.286	67.061	78.916	93.560	112.04	135.98
	.10	29.655	34.551	40.204	46.803	54.606	63.964	75.377	89.575	107.66
28	.01	50.604	58.756	68.291	79.582	93.138	109.67	130.20	156.26	190.16
	.05	35.043	40.624	47.053	54.538	63.357	73.890	86.668	102.45	122.38
	.10	28.821	33.427	38.694	44.778	51.883	60.286	70.367	82.663	97.960
29	.01	48.839	56.406	65.165	75.416	87.560	102.14	119.93	142.03	170.06
	.05	34.044	39.291	45.280	52.183	60.223	69.700	81.021	94.757	111.73
	.10	28.083	32.438	37.378	43.031	49.563	57.194	66.221	77.052	90.265
30	.01	47.283	54.353	62.461	71.851	82.847	95.877	111.53	130.62	154.34
	.05	33.156	38.115	43.730	50.143	57.539	66.156	76.316	88.455	103.18
	.10	27.424	31.563	36.222	41.509	47.563	54.560	62.737	72.412	84.020

TABLE A.18.   UPPER PERCENTAGE POINTS OF HOTELLING'S T^2 DISTRIBUTION (1-CDF)
              (cont.)

ν	α	p=10	11	12	13	14	15	16	17	18
35	.01	41.651	47.059	53.053	59.741	67.252	75.749	85.434	96.566	109.48
	.05	29.881	33.848	38.209	43.030	48.392	54.392	61.152	68.824	77.602
	.10	24.972	28.354	32.059	36.137	40.652	45.679	51.311	57.665	64.887
40	.01	38.135	42.617	47.478	52.776	58.578	64.961	72.020	79.868	88.644
	.05	27.783	31.175	34.833	38.794	43.102	47.807	52.969	58.658	64.961
	.10	23.382	26.318	29.475	32.885	36.582	40.605	45.003	49.831	55.156
45	.01	35.737	39.636	43.803	48.272	53.083	58.281	63.917	70.051	76.754
	.05	26.326	29.346	32.559	35.990	39.665	43.614	47.871	52.476	57.475
	.10	22.269	24.911	27.718	30.709	33.905	37.332	41.016	44.990	49.290
50	.01	33.998	37.501	41.203	45.128	49.301	53.752	58.511	63.617	69.109
	.05	25.256	28.017	30.926	34.000	37.256	40.715	44.398	48.330	52.540
	.10	21.446	23.882	26.446	29.151	32.013	35.047	38.272	41.708	45.377
55	.01	32.682	35.898	39.268	42.811	46.543	50.484	54.657	59.085	63.795
	.05	24.437	27.008	29.696	32.514	35.475	38.593	41.882	45.361	49.047
	.10	20.814	23.097	25.483	27.982	30.605	33.363	36.270	39.338	42.584
60	.01	31.650	34.650	37.774	41.034	44.444	48.019	51.774	55.726	59.894
	.05	23.790	26.216	28.737	31.364	34.106	36.973	39.978	43.131	46.447
	.10	20.312	22.478	24.729	27.073	29.517	32.072	34.745	37.548	40.491
70	.01	30.139	32.836	35.617	38.490	41.465	44.549	47.753	51.085	54.557
	.05	22.834	25.053	27.339	29.699	32.139	34.666	37.287	40.009	42.840
	.10	19.568	21.566	23.625	25.750	27.946	30.220	32.578	35.024	37.567
80	.01	29.085	31.581	34.137	36.759	39.453	42.226	45.085	48.035	51.084
	.05	22.162	24.241	26.370	28.553	30.796	33.103	35.479	37.929	40.457
	.10	19.042	20.925	22.855	24.834	26.867	28.958	31.112	33.331	35.622
90	.01	28.310	30.662	33.059	35.504	38.004	40.564	43.187	45.880	48.645
	.05	21.663	23.642	25.658	27.716	29.820	31.974	34.180	36.444	38.767
	.10	18.651	20.451	22.287	24.163	26.080	28.043	30.055	32.118	34.235

TABLE A.18.   UPPER PERCENTAGE POINTS OF HOTELLING'S T^2 DISTRIBUTION (1-CDF) (cont.)

ν	α	p=10	11	12	13	14	15	16	17	18
100	.01	27.714	29.960	32.238	34.554	36.912	39.316	41.769	44.276	46.840
	.05	21.279	23.182	25.114	27.079	29.080	31.120	33.203	35.331	37.507
	.10	18.348	20.086	21.852	23.649	25.481	27.349	29.256	31.205	33.197
110	.01	27.243	29.406	31.593	33.810	36.059	38.344	40.669	43.037	45.451
	.05	20.973	22.817	24.683	26.577	28.499	30.453	32.441	34.466	36.530
	.10	18.107	19.796	21.507	23.244	25.009	26.805	28.632	30.493	32.391
120	.01	26.862	28.958	31.073	33.210	35.374	37.567	39.792	42.051	44.348
	.05	20.725	22.521	24.335	26.171	28.030	29.916	31.830	33.775	35.752
	.10	17.911	19.560	21.228	22.917	24.629	26.366	28.131	29.924	31.747
150	.01	26.054	28.013	29.980	31.957	33.947	35.952	37.976	40.019	42.085
	.05	20.196	21.894	23.600	25.317	27.049	28.795	30.559	32.342	34.144
	.10	17.493	19.059	20.636	22.225	23.829	25.448	27.084	28.738	30.412
200	.01	25.287	27.122	28.953	30.784	32.619	34.457	36.303	38.157	40.021
	.05	19.692	21.297	22.904	24.514	26.128	27.749	29.378	31.017	32.665
	.10	17.092	18.582	20.075	21.572	23.076	24.587	26.108	27.638	29.179
400	.01	24.209	25.874	27.525	29.163	30.792	32.414	34.030	35.641	37.249
	.05	18.976	20.457	21.928	23.392	24.851	26.306	27.758	29.209	30.658
	.10	16.523	17.906	19.284	20.657	22.027	23.395	24.762	26.129	27.497
1000	.01	23.600	25.174	26.727	28.262	29.782	31.289	32.785	34.271	35.749
	.05	18.570	19.981	21.379	22.764	24.139	25.505	26.862	28.215	29.561
	.10	16.197	17.522	18.837	20.142	21.440	22.731	24.015	25.297	26.573
∞	.01	23.209	24.725	26.217	27.688	29.141	30.578	32.000	33.409	34.805
	.05	18.307	19.675	21.026	22.362	23.685	24.996	26.296	27.587	28.869
	.10	15.987	17.275	18.549	19.812	21.064	22.307	23.542	24.769	25.989

TABLE A.18.   UPPER PERCENTAGE POINTS OF HOTELLING'S T^2 DISTRIBUTION (1-CDF)
(cont.)

ν	α	p=20	22	24	26	28	30	35	40	50
22	.01	3914.5	---	---	---	---	---	---	---	---
	.05	1270.2	---	---	---	---	---	---	---	---
	.10	760.39	---	---	---	---	---	---	---	---
24	.01	917.05	4688.6	---	---	---	---	---	---	---
	.05	437.58	1522.1	---	---	---	---	---	---	---
	.10	307.84	911.69	---	---	---	---	---	---	---
26	.01	457.26	1087.5	5532.3	---	---	---	---	---	---
	.05	255.88	519.52	1796.8	---	---	---	---	---	---
	.10	192.75	365.84	1076.7	---	---	---	---	---	---
28	.01	299.16	537.80	1272.3	6445.7	---	---	---	---	---
	.05	182.71	301.49	608.45	2094.2	---	---	---	---	---
	.10	143.01	227.41	428.81	1255.4	---	---	---	---	---
30	.01	223.58	349.44	624.79	1471.6	7428.7	---	---	---	---
	.05	144.35	213.91	350.79	704.36	2414.4	---	---	---	---
	.10	115.80	167.69	264.89	496.74	1447.7	---	---	---	---
35	.01	142.57	190.46	264.63	390.97	639.63	1265.0	---	---	---
	.05	99.556	130.17	175.38	247.78	378.49	666.43	---	---	---
	.10	82.743	107.20	142.52	197.44	292.66	489.99	---	---	---
40	.01	109.70	137.13	174.13	226.27	304.12	429.94	1675.3	---	---
	.05	79.849	98.797	123.66	157.54	206.07	280.42	884.07	---	---
	.10	67.640	83.352	103.69	130.98	169.28	226.50	650.83	---	---
45	.01	92.215	111.18	134.94	165.47	205.92	261.62	557.66	2143.0	---
	.05	68.879	82.644	99.572	120.86	148.36	185.10	364.83	1132.3	---
	.10	59.050	70.739	84.987	102.72	125.35	155.15	295.28	834.35	---
50	.01	81.450	96.008	113.43	134.63	160.93	194.28	334.02	701.74	---
	.05	61.921	72.859	85.775	101.25	120.12	143.59	237.28	460.16	---
	.10	53.524	62.969	74.052	87.236	103.17	122.80	199.39	373.02	---
60	.01	68.958	79.158	90.728	103.97	119.27	137.14	198.22	301.24	1039.0
	.05	53.624	61.642	70.667	80.905	92.623	106.16	151.36	224.52	683.62
	.10	46.848	53.926	61.864	70.832	81.047	92.789	131.55	193.02	555.39

TABLE A.18.   UPPER PERCENTAGE POINTS OF HOTELLING'S T^2 DISTRIBUTION (1-CDF)
(cont.)

ν	α	p=20	22	24	26	28	30	35	40	50*
70	.01	61.964	70.070	78.992	88.868	99.865	112.19	150.64	205.48	430.64
	.05	48.859	55.417	62.599	70.504	79.253	88.990	118.95	160.68	322.67
	.10	42.966	48.838	55.253	62.296	70.067	78.688	105.04	141.31	278.33
80	.01	57.504	64.405	71.854	79.929	88.717	98.325	126.80	163.98	286.52
	.05	45.771	51.467	57.594	64.210	71.381	79.185	102.10	131.56	225.53
	.10	40.432	45.581	51.112	57.074	63.524	70.528	91.000	117.12	199.13
90	.01	54.417	60.542	67.065	74.034	81.504	89.536	112.57	141.05	224.42
	.05	43.610	48.739	54.190	59.997	66.204	72.857	91.809	114.99	181.39
	.10	38.648	43.319	48.277	53.554	59.187	65.215	82.336	103.17	162.20
100	.01	52.154	57.741	63.632	69.860	76.462	83.478	103.14	126.58	190.24
	.05	42.013	46.744	51.725	56.983	62.544	68.440	84.886	104.33	156.35
	.10	37.324	41.656	46.215	51.022	56.103	61.485	76.462	94.104	140.94
120	.01	49.063	53.955	59.044	64.347	69.887	75.682	91.448	109.38	153.97
	.05	39.811	44.022	48.399	52.958	57.715	62.685	76.172	91.440	129.08
	.10	35.492	39.378	43.417	47.624	52.011	56.594	69.015	83.048	117.50
150	.01	46.287	50.596	55.023	59.579	64.275	69.123	81.978	96.041	128.70
	.05	37.815	41.581	45.451	49.435	53.540	57.777	69.003	81.259	109.60
	.10	33.823	37.324	40.924	44.631	48.452	52.394	62.839	74.235	100.54
200	.01	43.783	47.599	51.475	55.419	59.438	63.536	74.171	85.434	110.19
	.05	35.997	39.381	42.822	46.327	49.900	53.545	63.009	73.029	95.040
	.10	32.296	35.464	38.690	41.977	45.329	48.751	57.640	67.055	87.730
400	.01	40.459	43.667	46.877	50.085	53.325	56.569	64.756	73.081	90.249
	.05	33.558	36.462	39.375	42.300	45.239	48.195	55.671	63.291	79.039
	.10	30.236	32.984	35.743	38.517	41.307	44.116	51.229	58.488	73.517
1000	.01	38.677	41.581	44.465	47.333	50.187	53.031	60.109	67.161	81.261
	.05	32.238	34.899	37.547	40.186	42.819	45.466	52.002	58.556	71.707
	.10	29.116	31.648	34.172	36.691	39.206	41.719	48.001	54.293	66.948
∞	.01	37.566	40.289	42.980	45.642	48.278	50.892	57.342	63.691	76.154
	.05	31.410	33.924	36.415	38.885	41.337	43.773	49.802	55.758	67.505
	.10	28.412	30.813	33.196	35.563	37.916	40.256	46.059	51.805	63.167

Source: Values were extracted, by permission, from tables produced by
D. R. Jensen and R. B. Howe, 1972, Tech. Rep. 9 (rev.), Dept. Stat. (Va. Polytech. Inst. and State Univ., Blacksburg).

*For $p > 50$ (or $\alpha > 0.10$), compare $f = [(\nu-p+1)/p\nu]T^2$ with $f_{\alpha,p,\nu-p+1}$
($\nu \geq p$), Table A.5.

TABLE A.19. LOWER PERCENTAGE POINTS FOR WILKS'S LIKELIHOOD-RATIO U TEST (CDF)*

ν_1	α	$\nu_2=4$	5	6	7	8	9	10	11	12
					$p=2$					
1	.05	.1357	.2236	.3017	.3684	.4249	.4729	.5139	.5493	.5800
	.01	.0464	.1000	.1585	.2154	.2683	.3162	.3594	.3981	.4329
2	.05	.0618	.1174	.1749	.2297	.2802	.3259	.3670	.4041	.4373
	.01	.0198	.0493	.0866	.1272	.1681	.2079	.2457	.2811	.3141
3	.05	.0358	.0736	.1164	.1602	.2028	.2432	.2808	.3157	.3480
	.01	.0112	.0300	.0558	.0860	.1181	.1507	.1829	.2141	.2439
4	.05	.0235	.0508	.0837	.1190	.1547	.1898	.2234	.2554	.2855
	.01	.0072	.0202	.0393	.0625	.0883	.1153	.1427	.1699	.1965
5	.05	.0166	.0372	.0632	.0921	.1224	.1528	.1826	.2116	.2394
	.01	.0050	.0146	.0292	.0477	.0688	.0914	.1150	.1388	.1625
6	.05	.0123	.0285	.0495	.0736	.0994	.1259	.1524	.1785	.2040
	.01	.0037	.0111	.0226	.0376	.0552	.0745	.0948	.1158	.1369
					$p=3$					
1	.05	.0337	.0974	.1683	.2355	.2960	.3493	.3961	.4371	.4734
	.01	.0068	.0329	.0740	.1214	.1698	.2164	.2600	.3002	.3372
2	.05	.0096	.0359	.0736	.1165	.1602	.2028	.2431	.2808	.3157
	.01	.0018	.0112	.0300	.0559	.0860	.1181	.1507	.1829	.2141
3	.05	.0042	.0175	.0397	.0677	.0989	.1314	.1638	.1956	.2261
	.01	.0008	.0053	.0155	.0312	.0510	.0737	.0980	.1232	.1485
4	.05	.0023	.0100	.0240	.0432	.0659	.0908	.1167	.1429	.1689
	.01	.0005	.0030	.0092	.0194	.0331	.0496	.0681	.0879	.1084
5	.05	.0015	.0064	.0158	.0294	.0464	.0657	.0864	.1081	.1301
	.01	.0003	.0020	.0060	.0130	.0229	.0352	.0495	.0652	.0820
6	.05	.0011	.0044	.0110	.0210	.0340	.0492	.0660	.0840	.1026
	.01	.0003	.0014	.0042	.0092	.0166	.0260	.0373	.0500	.0638

*Values tabulated are useful in testing p-dimensional vectors of means in multivariate analysis of variance. The hypothesis in question is rejected if the computed U statistic is *smaller* than the tabular value. If required values of p and ν_1 exceed those provided in the table, transformations of the U statistic may be made to permit approximate evaluation by other tables (see Chap. 4). See Kramer (1972, *A first course in methods of multivariate analysis*, Va. Polytech. Inst. & State Univ., Dept. Stat., Blacksburg), however, for $p \leq 8$, $\nu_1 \leq 120$.

TABLE A.19. LOWER PERCENTAGE POINTS FOR WILKS'S LIKELIHOOD-RATIO U TEST (CDF) (<u>cont.</u>)

ν_1	α	$\nu_2=4$	5	6	7	8	9	10	11	12
					p=4					
1	.05	.1378*	.0255	.0761	.1354	.1940	.2486	.2981	.3426	.3824
	.01	.0090*	.0052	.0256	.0590	.0989	.1409	.1824	.2221	.2595
2	.05	.0292*	.0061	.0236	.0508	.0837	.1190	.1548	.1898	.2234
	.01	.0026*	.0012	.0073	.0204	.0393	.0626	.0883	.1153	.1427
3	.05	.0127*	.0023	.0100	.0240	.0432	.0659	.0908	.1167	.1429
	.01	.0015*	.0005	.0030	.0092	.0194	.0331	.0496	.0681	.0879
4	.05	.0075*	.0011	.0051	.0130	.0249	.0399	.0574	.0765	.0967
	.01	.0011*	.0002	.0015	.0049	.0109	.0195	.0304	.0434	.0578
5	.05	.0052*	.0006	.0029	.0077	.0154	.0257	.0383	.0525	.0681
	.01	.0009*	.0002	.0009	.0029	.0066	.0123	.0199	.0291	.0398
6	.05	.0040*	.0004	.0018	.0049	.0101	.0174	.0266	.0374	.0495
	.01	.0008*	.0001	.0006	.0018	.0043	.0082	.0136	.0204	.0285
					p=5					
1	.05	...	.1598*	.0211	.0628	.1135	.1654	.2148	.2606	.3026
	.01	...	.0164*	.0047	.0213	.0493	.0837	.1207	.1580	.1944
2	.05	...	.0291*	.0044	.0169	.0374	.0633	.0922	.1224	.1528
	.01	...	.0036*	.0010	.0053	.0149	.0294	.0478	.0688	.0915
3	.05	...	.0105*	.0015	.0064	.0158	.0294	.0464	.0657	.0864
	.01	...	.0015*	.0003	.0020	.0060	.0130	.0229	.0352	.0495
4	.05	...	.0052*	.0006	.0029	.0077	.0154	.0257	.0383	.0525
	.01	...	.0009*	.0002	.0009	.0029	.0066	.0123	.0199	.0291
5	.05	...	.0031*	.0003	.0015	.0042	.0088	.0153	.0237	.0336
	.01	...	.0006*	.0001	.0005	.0016	.0037	.0072	.0120	.0182
6	.05	...	.0021*	.0002*	.0009	.0025	.0053	.0096	.0154	.0224
	.01	...	.0004*	.0052*	.0003	.0009	.0022	.0044	.0077	.0119

*Divide these entries by 100.

TABLE A.19.   LOWER PERCENTAGE POINTS FOR WILKS'S LIKELIHOOD-RATIO U TEST (CDF) (cont.)

ν_1	α	$\nu_2=7$	8	9	10	11	12	13	14
					p=6				
1	.05	.0188	.0539	.0980	.1443	.1894	.2319	.2714	.3078
	.01	.0046	.0188	.0428	.0729	.1059	.1397	.1732	.2055
2	.05	.0035	.0129	.0288	.0497	.0737	.0994	.1259	.1525
	.01	.0008	.0042	.0115	.0229	.0378	.0553	.0746	.0949
3	.05	.0011	.0044	.0110	.0210	.0340	.0492	.0660	.0840
	.01	.0003	.0014	.0042	.0092	.0166	.0260	.0373	.0500
4	.05	.0004	.0018	.0049	.0101	.0174	.0266	.0374	.0495
	.01	.0001	.0006	.0018	.0043	.0082	.0136	.0204	.0285
5	.05	.0002 *	.0009	.0025	.0053	.0096	.0154	.0224	.0307
	.01	.0052	.0003	.0009	.0022	.0044	.0077	.0119	.0172
6	.05	.0001 *	.0005	.0014	.0030	.0057	.0093	.0141	.0198
	.01	.0029	.0002	.0005	.0013	.0026	.0046	.0073	.0109
					p=7				
1	.05	.2625 * *	.0176 * 	.0478	.0866	.1282	.1695	.2090	.2462
	.01	.0486	.4798	.0173	.0382	.0648	.0946	.1254	.1563
2	.05	.0350 * *	.0030 * 	.0103	.0231	.0402	.0604	.0825	.1057
	.01	.0068	.0782	.0035	.0093	.0186	.0309	.0456	.0621
3	.05	.0091 * *	.0008 * 	.0032	.0081	.0156	.0257	.0379	.0516
	.01	.0019	.0215	.0010	.0031	.0069	.0125	.0198	.0288
4	.05	.0033 * *	.0003 * 	.0012	.0033	.0070	.0122	.0191	.0274
	.01	.0007	.0079	.0004	.0013	.0030	.0057	.0097	.0148
5	.05	.0015 * *	.0001 * 	.0005	.0016	.0034	.0063	.0104	.0155
	.01	.0003	.0035	.0002	.0006	.0014	.0029	.0051	.0081
6	.05	.0008 * *	.0063 * *	.0003	.0008	.0018	.0035	.0059	.0092
	.01	.0002	.0018	.0001	.0003	.0008	.0016	.0029	.0047

*Divide these entries by 100.

TABLE A.19.   LOWER PERCENTAGE POINTS FOR WILKS'S LIKELIHOOD-RATIO U TEST (CDF)
(cont.)

ν_1	α	ν_2=13	14	15	16	17	18	19	20	22
					p=2					
1	.05	.6070	.6307	.6519	.6707	.6877	.7030	.7169	.7295	.7518
	.01	.4642	.4924	.5179	.5412	.5623	.5817	.5995	.6158	.6449
2	.05	.4674	.4946	.5193	.5418	.5623	.5811	.5985	.6145	.6430
	.01	.3448	.3732	.3996	.4240	.4467	.4678	.4875	.5058	.5390
3	.05	.3777	.4052	.4306	.4540	.4757	.4959	.5146	.5321	.5636
	.01	.2722	.2990	.3243	.3482	.3707	.3918	.4118	.4305	.4650
4	.05	.3138	.3404	.3653	.3885	.4103	.4308	.4500	.4680	.5009
	.01	.2223	.2471	.2708	.2934	.3150	.3356	.3551	.3737	.4081
5	.05	.2658	.2910	.3149	.3374	.3588	.3790	.3980	.4161	.4494
	.01	.1858	.2085	.2305	.2518	.2722	.2918	.3107	.3287	.3625
6	.05	.2286	.2522	.2748	.2964	.3170	.3363	.3553	.3732	.4063
	.01	.1580	.1788	.1991	.2189	.2382	.2568	.2748	.2921	.3249
					p=3					
1	.05	.5055	.5340	.5596	.5826	.6033	.6222	.6393	.6550	.6827
	.01	.3710	.4019	.4302	.4561	.4800	.5019	.5222	.5409	.5745
2	.05	.3480	.3777	.4052	.4306	.4540	.4757	.4959	.5146	.5484
	.01	.2439	.2722	.2990	.3243	.3482	.3707	.3918	.4117	.4482
3	.05	.2552	.2828	.3090	.3336	.3568	.3786	.3992	.4186	.4542
	.01	.1735	.1980	.2218	.2448	.2669	.2881	.3083	.3276	.3638
4	.05	.1944	.2191	.2429	.2658	.2877	.3086	.3286	.3475	.3829
	.01	.1293	.1502	.1709	.1913	.2112	.2305	.2493	.2675	.3019
5	.05	.1522	.1740	.1953	.2161	.2363	.2559	.2747	.2928	.3271
	.01	.0994	.1172	.1352	.1531	.1708	.1883	.2055	.2223	.2546
6	.05	.1217	.1408	.1598	.1786	.1970	.2150	.2326	.2497	.2823
	.01	.0784	.0936	.1091	.1249	.1406	.1564	.1720	.1873	.2173

TABLE A.19. LOWER PERCENTAGE POINTS FOR WILKS'S LIKELIHOOD-RATIO U TEST (CDF)
            (cont.)

ν_1	α	ν_2=13	14	15	16	17	18	19	20	22
					p=4					
1	.05	.4182	.4503	.4793	.5055	.5293	.5510	.5709	.5891	.6213
	.01	.2943	.3267	.3566	.3844	.4101	.4339	.4560	.4765	.5135
2	.05	.2554	.2855	.3138	.3404	.3653	.3885	.4103	.4308	.4680
	.01	.1699	.1965	.2223	.2471	.2708	.2934	.3150	.3356	.3737
3	.05	.1689	.1944	.2191	.2429	.2658	.2877	.3086	.3286	.3657
	.01	.1084	.1293	.1502	.1709	.1913	.2112	.2305	.2493	.2850
4	.05	.1174	.1383	.1591	.1797	.1998	.2195	.2386	.2571	.2921
	.01	.0733	.0896	.1064	.1234	.1406	.1576	.1745	.1911	.2234
5	.05	.0845	.1016	.1190	.1364	.1539	.1712	.1882	.2049	.2372
	.01	.0517	.0645	.0780	.0920	.1063	.1208	.1353	.1499	.1787
6	.05	.0626	.0765	.0910	.1058	.1208	.1359	.1509	.1659	.1952
	.01	.0377	.0478	.0587	.0702	.0822	.0945	.1070	.1197	.1451
					p=5					
1	.05	.3408	.3755	.4071	.4359	.4622	.4863	.5084	.5287	.5649
	.01	.2291	.2619	.2927	.3215	.3484	.3736	.3971	.4190	.4587
2	.05	.1827	.2116	.2394	.2659	.2910	.3149	.3374	.3588	.3980
	.01	.1150	.1388	.1625	.1858	.2085	.2305	.2518	.2722	.3107
3	.05	.1081	.1301	.1522	.1740	.1953	.2161	.2363	.2559	.2928
	.01	.0652	.0820	.0994	.1172	.1352	.1531	.1708	.1883	.2223
4	.05	.0681	.0845	.1016	.1190	.1364	.1539	.1712	.1882	.2213
	.01	.0398	.0517	.0645	.0780	.0920	.1063	.1208	.1353	.1644
5	.05	.0449	.0572	.0703	.0840	.0982	.1126	.1271	.1417	.1706
	.01	.0256	.0342	.0437	.0539	.0648	.0762	.0879	.1000	.1245
6	.05	.0307	.0400	.0501	.0610	.0724	.0842	.0963	.1086	.1336
	.01	.0172	.0235	.0306	.0384	.0470	.0560	.0656	.0755	.0960

TABLE A.19.   LOWER PERCENTAGE POINTS FOR WILKS'S LIKELIHOOD-RATIO U TEST (CDF)
(cont.)

ν_1	α	ν_2=15	16	17	18	19	20	22	24
					p=6				
1	.05	.3413	.3720	.4003	.4264	.4503	.4726	.5122	.5465
	.01	.2364	.2656	.2932	.3192	.3437	.3667	.4086	.4457
2	.05	.1786	.2040	.2286	.2522	.2748	.2964	.3366	.3731
	.01	.1158	.1370	.1580	.1788	.1991	.2189	.2568	.2921
3	.05	.1026	.1217	.1408	.1598	.1786	.1970	.2326	.2662
	.01	.0638	.0784	.0936	.1091	.1249	.1406	.1720	.2025
4	.05	.0626	.0765	.0910	.1058	.1208	.1359	.1659	.1952
	.01	.0377	.0478	.0587	.0702	.0822	.0945	.1197	.1451
5	.05	.0400	.0501	.0610	.0724	.0842	.0963	.1211	.1461
	.01	.0235	.0306	.0384	.0470	.0560	.0656	.0856	.1066
6	.05	.0264	.0339	.0421	.0508	.0601	.0698	.0901	.1112
	.01	.0152	.0203	.0260	.0324	.0393	.0467	.0627	.0799
					p=7				
1	.05	.2809	.3130	.3428	.3705	.3960	.4198	.4624	.4995
	.01	.1865	.2156	.2434	.2698	.2949	.3186	.3621	.4010
2	.05	.1293	.1529	.1762	.1989	.2209	.2423	.2825	.3196
	.01	.0798	.0983	.1172	.1363	.1552	.1740	.2105	.2452
3	.05	.0667	.0825	.0990	.1157	.1326	.1495	.1828	.2149
	.01	.0392	.0507	.0630	.0761	.0896	.1034	.1315	.1597
4	.05	.0369	.0475	.0589	.0709	.0834	.0963	.1226	.1492
	.01	.0210	.0282	.0363	.0452	.0547	.0647	.0859	.1081
5	.05	.0216	.0287	.0366	.0452	.0544	.0641	.0846	.1061
	.01	.0119	.0166	.0220	.0281	.0348	.0421	.0580	.0753
6	.05	.0132	.0180	.0235	.0297	.0365	.0438	.0597	.0769
	.01	.0072	.0102	.0139	.0181	.0229	.0282	.0402	.0536

TABLE A.19. LOWER PERCENTAGE POINTS FOR WILKS'S LIKELIHOOD-RATIO U TEST (CDF)
(<u>cont.</u>)

ν_1	α	$\nu_2=24$	26	28	30	40	60	80	100	120
					$p=2$					
1	.05	.7707	.7869	.8010	.8133	.8576	.9034	.9270	.9413	.9509
	.01	.6700	.6918	.7110	.7279	.7897	.8555	.8900	.9112	.9255
2	.05	.6677	.6892	.7081	.7249	.7864	.8526	.8875	.9091	.9237
	.01	.5681	.5939	.6169	.6375	.7145	.8000	.8462	.8751	.8948
3	.05	.5913	.6158	.6375	.6570	.7298	.8107	.8543	.8817	.9004
	.01	.4957	.5233	.5481	.5706	.6567	.7556	.8104	.8452	.8693
4	.05	.5301	.5563	.5797	.6009	.6816	.7738	.8247	.8570	.8792
	.01	.4393	.4676	.4933	.5168	.6086	.7173	.7791	.8188	.8464
5	.05	.4794	.5064	.5309	.5532	.6394	.7406	.7976	.8342	.8596
	.01	.3935	.4219	.4479	.4719	.5672	.6833	.7508	.7946	.8254
6	.05	.4364	.4638	.4888	.5117	.6019	.7102	.7725	.8128	.8411
	.01	.3553	.3834	.4094	.4335	.5308	.6526	.7248	.7723	.8059
					$p=3$					
1	.05	.7063	.7267	.7444	.7600	.8161	.8748	.9052	.9237	.9362
	.01	.6036	.6290	.6515	.6714	.7447	.8237	.8654	.8912	.9087
2	.05	.5779	.6039	.6269	.6475	.7239	.8078	.8527	.8806	.8996
	.01	.4808	.5099	.5360	.5597	.6496	.7520	.8083	.8438	.8682
3	.05	.4859	.5143	.5399	.5630	.6514	.7524	.8083	.8436	.8680
	.01	.3967	.4267	.4540	.4791	.5775	.6947	.7614	.8043	.8342
4	.05	.4151	.4443	.4710	.4953	.5908	.7042	.7688	.8103	.8393
	.01	.3339	.3635	.3909	.4163	.5187	.6458	.7205	.7693	.8037
5	.05	.3587	.3878	.4147	.4395	.5388	.6613	.7330	.7797	.8126
	.01	.2850	.3136	.3405	.3656	.4693	.6030	.6838	.7376	.7758
6	.05	.3127	.3412	.3677	.3925	.4937	.6226	.7000	.7513	.7877
	.01	.2460	.2733	.2992	.3236	.4269	.5648	.6505	.7084	.7500

TABLE A.19.   LOWER PERCENTAGE POINTS FOR WILKS'S LIKELIHOOD-RATIO U TEST (CDF) (<u>cont.</u>)

ν_1	α	$\nu_2=24$	26	28	30	40	60	80	100	120
					$p=4$					
1	.05	.6489	.6729	.6938	.7122	.7789	.8490	.8854	.9077	.9227
	.01	.5458	.5742	.5994	.6218	.7048	.7953	.8434	.8733	.8936
2	.05	.5009	.5301	.5563	.5797	.6682	.7670	.8207	.8543	.8773
	.01	.4081	.4393	.4676	.4933	.5930	.7092	.7741	.8155	.8440
3	.05	.3994	.4300	.4579	.4834	.5828	.7001	.7663	.8086	.8380
	.01	.3182	.3490	.3775	.4039	.5100	.6410	.7175	.7673	.8022
4	.05	.3248	.3550	.3830	.4089	.5133	.6426	.7183	.7677	.8024
	.01	.2542	.2834	.3108	.3367	.4441	.5837	.6685	.7249	.7650
5	.05	.2679	.2968	.3240	.3495	.4552	.5921	.6751	.7304	.7696
	.01	.2067	.2337	.2596	.2842	.3900	.5343	.6251	.6867	.7311
6	.05	.2235	.2507	.2766	.3012	.4059	.5473	.6359	.6959	.7391
	.01	.1704	.1951	.2192	.2424	.3449	.4909	.5861	.6519	.6999
					$p=5$					
1	.05	.5960	.6231	.6468	.6678	.7440	.8248	.8668	.8926	.9101
	.01	.4937	.5246	.5520	.5766	.6682	.7691	.8230	.8566	.8795
2	.05	.4332	.4648	.4932	.5189	.6172	.7292	.7907	.8296	.8563
	.01	.3460	.3783	.4080	.4352	.5423	.6700	.7425	.7891	.8214
3	.05	.3271	.3587	.3878	.4147	.5217	.6520	.7272	.7758	.8098
	.01	.2546	.2850	.3136	.3405	.4511	.5926	.6772	.7330	.7725
4	.05	.2528	.2826	.3106	.3369	.4460	.5869	.6718	.7280	.7680
	.01	.1928	.2203	.2468	.2721	.3807	.5286	.6214	.6841	.7292
5	.05	.1988	.2261	.2522	.2772	.3844	.5307	.6225	.6848	.7297
	.01	.1491	.1736	.1975	.2207	.3245	.4743	.5724	.6404	.6901
6	.05	.1585	.1831	.2071	.2303	.3335	.4816	.5783	.6453	.6943
	.01	.1173	.1387	.1601	.1812	.2788	.4276	.5290	.6010	.6542

TABLE A.19. LOWER PERCENTAGE POINTS FOR WILKS'S LIKELIHOOD-RATIO U TEST (CDF) (<u>cont.</u>)

ν_1	α	ν_2=26	28	30	40	60	80	100	120
					p=6				
1	.05	.5763	.6026	.6259	.7109	.8016	.8491	.8782	.8979
	.01	.4787	.5082	.5346	.6340	.7473	.8037	.8408	.8661
2	.05	.4063	.4364	.4638	.5700	.6935	.7623	.8059	.8361
	.01	.3249	.3553	.3834	.4960	.6335	.7128	.7641	.8000
3	.05	.2977	.3272	.3547	.4668	.6075	.6905	.7447	.7829
	.01	.2318	.2598	.2864	.3990	.5483	.6398	.7009	.7445
4	.05	.2235	.2507	.2766	.3872	.5362	.6286	.6908	.7354
	.01	.1704	.1951	.2192	.3262	.4791	.5781	.6462	.6957
5	.05	.1709	.1952	.2188	.3242	.4756	.5743	.6425	.6921
	.01	.1280	.1495	.1707	.2698	.4214	.5248	.5979	.6519
6	.05	.1326	.1540	.1752	.2735	.4237	.5262	.5988	.6525
	.01	.0978	.1163	.1349	.2252	.3726	.4780	.5546	.6121
					p=7				
1	.05	.5319	.5606	.5860	.6792	.7793	.8319	.8643	.8862
	.01	.4358	.4670	.4951	.6015	.7207	.7853	.8257	.8533
2	.05	.3536	.3848	.4134	.5260	.6596	.7350	.7833	.8167
	.01	.2778	.3084	.3370	.4535	.5992	.6847	.7404	.7794
3	.05	.2456	.2748	.3022	.4170	.5660	.6558	.7151	.7572
	.01	.1873	.2142	.2400	.3525	.5075	.6048	.6706	.7179
4	.05	.1754	.2011	.2259	.3354	.4897	.5883	.6557	.7044
	.01	.1307	.1533	.1757	.2791	.4343	.5382	.6108	.6641
5	.05	.1280	.1499	.1717	.2727	.4261	.5299	.6029	.6567
	.01	.0935	.1123	.1313	.2238	.3744	.4813	.5584	.6162
6	.05	.0949	.1135	.1322	.2236	.3726	.4787	.5557	.6134
	.01	.0683	.0837	.0996	.1814	.3247	.4321	.5121	.5731

TABLE A.19. LOWER PERCENTAGE POINTS FOR WILKS'S LIKELIHOOD-RATIO U TEST (CDF) (cont.)

ν_1	α	ν_2=140	170	200	240	320	440	600	800	1000
						p=2				
1	.05	.9578	.9652	.9703	.9752	.9814	.9864	.9900	.9925	.9940
	.01	.9359	.9470	.9548	.9622	.9715	.9792	.9847	.9885	.9908
2	.05	.9342	.9456	.9536	.9612	.9707	.9786	.9843	.9882	.9906
	.01	.9092	.9247	.9356	.9461	.9593	.9702	.9781	.9835	.9868
3	.05	.9140	.9286	.9390	.9489	.9614	.9718	.9792	.9844	.9875
	.01	.8868	.9058	.9194	.9323	.9488	.9625	.9724	.9792	.9833
4	.05	.8955	.9131	.9256	.9376	.9528	.9654	.9745	.9808	.9846
	.01	.8668	.8888	.9047	.9199	.9392	.9554	.9671	.9752	.9801
5	.05	.8782	.8985	.9129	.9268	.9446	.9593	.9700	.9774	.9819
	.01	.8482	.8731	.8909	.9082	.9302	.9487	.9621	.9714	.9771
6	.05	.8619	.8846	.9009	.9166	.9367	.9535	.9656	.9741	.9792
	.01	.8308	.8582	.8780	.8971	.9217	.9423	.9574	.9678	.9742
						p=3				
1	.05	.9451	.9547	.9614	.9678	.9758	.9823	.9870	.9903	.9922
	.01	.9214	.9349	.9444	.9535	.9650	.9745	.9812	.9859	.9887
2	.05	.9134	.9282	.9387	.9487	.9613	.9717	.9792	.9844	.9875
	.01	.8861	.9053	.9190	.9321	.9487	.9624	.9723	.9792	.9833
3	.05	.8858	.9050	.9187	.9318	.9484	.9622	.9722	.9791	.9832
	.01	.8561	.8800	.8971	.9135	.9344	.9519	.9645	.9733	.9786
4	.05	.8605	.8837	.9002	.9161	.9364	.9533	.9656	.9741	.9792
	.01	.8292	.8571	.8771	.8965	.9213	.9422	.9572	.9678	.9741
5	.05	.8370	.8636	.8828	.9013	.9250	.9448	.9592	.9693	.9753
	.01	.8044	.8358	.8585	.8806	.9090	.9329	.9504	.9625	.9699
6	.05	.8148	.8446	.8662	.8871	.9140	.9366	.9531	.9646	.9716
	.01	.7812	.8158	.8410	.8655	.8972	.9241	.9437	.9575	.9658

TABLE A.19.   LOWER PERCENTAGE POINTS FOR WILKS'S LIKELIHOOD-RATIO U TEST (CDF)
(<u>cont.</u>)

ν_1	α	ν_2=140	170	200	240	320	440	600	800	1000
					p=4					
1	.05	.9336	.9451	.9532	.9609	.9706	.9786	.9843	.9882	.9905
	.01	.9083	.9240	.9351	.9457	.9591	.9701	.9780	.9835	.9868
2	.05	.8941	.9121	.9248	.9370	.9525	.9653	.9744	.9808	.9846
	.01	.8650	.8876	.9037	.9192	.9389	.9552	.9670	.9752	.9801
3	.05	.8596	.8830	.8997	.9158	.9362	.9532	.9655	.9740	.9792
	.01	.8281	.8563	.8766	.8961	.9211	.9420	.9572	.9677	.9741
4	.05	.8282	.8563	.8765	.8960	.9210	.9419	.9571	.9676	.9740
	.01	.7950	.8279	.8518	.8749	.9047	.9298	.9480	.9608	.9685
5	.05	.7990	.8313	.8546	.8773	.9065	.9311	.9490	.9615	.9691
	.01	.7646	.8017	.8287	.8551	.8892	.9182	.9393	.9541	.9631
6	.05	.7716	.8077	.8339	.8595	.8926	.9207	.9412	.9555	.9643
	.01	.7364	.7771	.8070	.8363	.8745	.9070	.9309	.9477	.9580
					p=5					
1	.05	.9226	.9360	.9455	.9545	.9657	.9750	.9816	.9862	.9890
	.01	.8961	.9139	.9265	.9385	.9536	.9661	.9751	.9813	.9850
2	.05	.8757	.8967	.9117	.9260	.9441	.9591	.9698	.9773	.9818
	.01	.8452	.8710	.8894	.9071	.9296	.9484	.9619	.9713	.9770
3	.05	.8348	.8621	.8817	.9005	.9245	.9446	.9591	.9692	.9753
	.01	.8019	.8340	.8572	.8797	.9085	.9326	.9502	.9624	.9698
4	.05	.7977	.8304	.8540	.8768	.9062	.9309	.9489	.9614	.9690
	.01	.7631	.8007	.8280	.8545	.8889	.9180	.9392	.9541	.9631
5	.05	.7634	.8008	.8280	.8545	.8888	.9179	.9391	.9540	.9630
	.01	.7278	.7699	.8008	.8311	.8705	.9041	.9287	.9461	.9566
6	.05	.7314	.7730	.8034	.8333	.8721	.9053	.9296	.9468	.9572
	.01	.6952	.7412	.7753	.8089	.8530	.8908	.9187	.9384	.9504

TABLE A.19. LOWER PERCENTAGE POINTS FOR WILKS'S LIKELIHOOD-RATIO U TEST (CDF)
            (cont.)

ν_1	α	ν_2=140	170	200	240	320	440	600	800	1000
					p=6					
1	.05	.9122	.9274	.9381	.9483	.9611	.9716	.9791	.9843	.9874
	.01	.8845	.9042	.9182	.9315	.9483	.9623	.9722	.9791	.9833
2	.05	.8582	.8820	.8990	.9153	.9359	.9531	.9654	.9740	.9791
	.01	.8264	.8551	.8757	.8955	.9207	.9418	.9571	.9677	.9741
3	.05	.8112	.8421	.8643	.8858	.9132	.9362	.9529	.9645	.9715
	.01	.7770	.8129	.8388	.8639	.8963	.9236	.9435	.9573	.9657
4	.05	.7688	.8056	.8324	.8584	.8920	.9203	.9410	.9554	.9642
	.01	.7331	.7748	.8053	.8351	.8738	.9066	.9307	.9476	.9579
5	.05	.7298	.7718	.8025	.8326	.8718	.9051	.9295	.9467	.9571
	.01	.6933	.7399	.7743	.8082	.8525	.8905	.9185	.9383	.9503
6	.05	.6937	.7401	.7744	.8082	.8525	.8904	.9184	.9382	.9503
	.01	.6568	.7075	.7453	.7828	.8323	.8751	.9068	.9293	.9430
					p=7					
1	.05	.9021	.9190	.9309	.9422	.9565	.9683	.9767	.9825	.9860
	.01	.8734	.8950	.9102	.9248	.9433	.9586	.9695	.9771	.9816
2	.05	.8412	.8678	.8867	.9049	.9280	.9472	.9611	.9707	.9765
	.01	.8083	.8398	.8625	.8843	.9122	.9355	.9524	.9641	.9712
3	.05	.7885	.8228	.8475	.8715	.9022	.9280	.9468	.9599	.9678
	.01	.7533	.7926	.8211	.8488	.8846	.9149	.9369	.9524	.9617
4	.05	.7411	.7818	.8116	.8405	.8781	.9099	.9332	.9495	.9594
	.01	.7047	.7502	.7836	.8164	.8591	.8956	.9224	.9413	.9528
5	.05	.6979	.7441	.7781	.8115	.8552	.8926	.9202	.9395	.9513
	.01	.6609	.7115	.7490	.7862	.8352	.8774	.9086	.9307	.9442
6	.05	.6581	.7089	.7467	.7841	.8334	.8760	.9075	.9298	.9435
	.01	.6209	.6758	.7169	.7580	.8125	.8600	.8953	.9205	.9359

TABLE A.19. LOWER PERCENTAGE POINTS FOR WILKS'S LIKELIHOOD-RATIO U TEST (CDF)
(cont.)

ν_1	α	ν_2=12	14	16	18	20	22	24	26
					p=2				
7	.05	.1762	.2209	.2628	.3014	.3370	.3696	.3994	.4269
	.01	.1172	.1552	.1924	.2281	.2618	.2934	.3229	.3505
8	.05	.1538	.1953	.2348	.2717	.3061	.3379	.3673	.3945
	.01	.1015	.1362	.1707	.2042	.2362	.2665	.2951	.3221
9	.05	.1355	.1741	.2112	.2464	.2794	.3103	.3391	.3659
	.01	.0888	.1206	.1526	.1840	.2144	.2434	.2710	.2972
10	.05	.1204	.1561	.1911	.2245	.2562	.2861	.3142	.3405
	.01	.0784	.1076	.1373	.1668	.1955	.2233	.2499	.2752
11	.05	.1077	.1409	.1738	.2055	.2359	.2648	.2921	.3178
	.01	.0698	.0966	.1242	.1519	.1792	.2057	.2313	.2558
12	.05	.0969	.1278	.1587	.1889	.2180	.2458	.2723	.2974
	.01	.0625	.0872	.1130	.1390	.1649	.1902	.2147	.2384
					p=3				
7	.05	.0825	.1157	.1495	.1828	.2149	.2456	.2747	.3022
	.01	.0507	.0761	.1034	.1315	.1597	.1873	.2142	.2400
8	.05	.0674	.0964	.1266	.1568	.1865	.2153	.2429	.2693
	.01	.0410	.0628	.0867	.1119	.1374	.1629	.1879	.2123
9	.05	.0559	.0812	.1082	.1357	.1631	.1900	.2160	.2411
	.01	.0338	.0525	.0735	.0960	.1193	.1427	.1659	.1888
10	.05	.0469	.0692	.0933	.1183	.1435	.1686	.1931	.2169
	.01	.0282	.0444	.0630	.0831	.1042	.1258	.1474	.1688
11	.05	.0398	.0594	.0810	.1038	.1270	.1503	.1734	.1959
	.01	.0238	.0379	.0544	.0725	.0917	.1115	.1316	.1517
12	.05	.0340	.0514	.0709	.0916	.1130	.1347	.1563	.1777
	.01	.0203	.0327	.0473	.0637	.0812	.0994	.1181	.1369

TABLE A.19.   LOWER PERCENTAGE POINTS FOR WILKS'S LIKELIHOOD-RATIO U TEST (CDF)
(<u>cont</u>.)

ν_1	α	ν_2=12	14	16	18	20	22	24	26
					p=4				
7	.05	.0369	.0589	.0834	.1094	.1359	.1624	.1883	.2136
	.01	.0210	.0363	.0547	.0751	.0969	.1194	.1420	.1646
8	.05	.0282	.0461	.0668	.0892	.1126	.1363	.1600	.1833
	.01	.0158	.0281	.0433	.0606	.0795	.0993	.1196	.1400
9	.05	.0219	.0367	.0542	.0736	.0942	.1154	.1370	.1584
	.01	.0122	.0222	.0348	.0495	.0659	.0833	.1015	.1200
10	.05	.0173	.0296	.0444	.0613	.0794	.0985	.1180	.1377
	.01	.0096	.0177	.0283	.0409	.0551	.0705	.0868	.1036
11	.05	.0139	.0241	.0368	.0515	.0675	.0846	.1023	.1203
	.01	.0077	.0144	.0233	.0341	.0465	.0602	.0747	.0899
12	.05	.0113	.0199	.0308	.0436	.0578	.0731	.0891	.1056
	.01	.0062	.0118	.0194	.0287	.0396	.0517	.0647	.0785
					p=5				
7	.05	.0155	.0287	.0452	.0641	.0846	.1061	.1279	.1499
	.01	.0081	.0166	.0281	.0421	.0580	.0753	.0935	.1123
8	.05	.0110	.0211	.0341	.0496	.0668	.0852	.1043	.1238
	.01	.0057	.0120	.0210	.0322	.0453	.0598	.0754	.0918
9	.05	.0080	.0158	.0262	.0389	.0534	.0691	.0858	.1031
	.01	.0041	.0089	.0159	.0250	.0358	.0481	.0615	.0758
10	.05	.0059	.0120	.0204	.0309	.0431	.0567	.0712	.0865
	.01	.0031	.0068	.0123	.0197	.0286	.0390	.0506	.0631
11	.05	.0045	.0093	.0161	.0248	.0351	.0468	.0595	.0731
	.01	.0023	.0052	.0096	.0157	.0232	.0320	.0420	.0529
12	.05	.0035	.0073	.0129	.0201	.0289	.0390	.0501	.0621
	.01	.0018	.0041	.0077	.0126	.0189	.0265	.0351	.0446

TABLE A.19.   LOWER PERCENTAGE POINTS FOR WILKS'S LIKELIHOOD-RATIO U TEST (CDF)
              (cont.)

ν_1	α	ν_2=12	14	16	18	20	22	24	26
					p=6				
7	.05	.0059	.0132	.0235	.0365	.0516	.0682	.0858	.1042
	.01	.0029	.0072	.0139	.0229	.0340	.0468	.0608	.0759
8	.05	.0039	.0091	.0167	.0267	.0387	.0523	.0671	.0827
	.01	.0019	.0049	.0097	.0165	.0252	.0354	.0470	.0596
9	.05	.0027	.0064	.0121	.0199	.0295	.0406	.0530	.0664
	.01	.0013	.0034	.0070	.0122	.0190	.0272	.0367	.0473
10	.05	.0019	.0046	.0090	.0151	.0228	.0319	.0423	.0537
	.01	.0009	.0024	.0051	.0091	.0145	.0212	.0291	.0379
11	.05	.0013	.0034	.0067	.0116	.0178	.0254	.0341	.0438
	.01	.0006	.0018	.0038	.0070	.0112	.0167	.0232	.0307
12	.05	.0010	.0025	.0051	.0090	.0140	.0203	.0277	.0360
	.01	.0005	.0013	.0029	.0054	.0088	.0133	.0187	.0250
					p=7				
7	.05	.0020	.0057	.0117	.0201	.0306	.0430	.0567	.0715
	.01	.0009	.0029	.0065	.0120	.0194	.0285	.0390	.0507
8	.05	.0013	.0036	.0078	.0139	.0218	.0314	.0424	.0545
	.01	.0006	.0018	.0043	.0082	.0136	.0205	.0288	.0382
9	.05	.0008	.0024	.0053	.0098	.0158	.0233	.0321	.0421
	.01	.0004	.0012	.0029	.0057	.0098	.0151	.0216	.0291
10	.05	.0005	.0016	.0037	.0071	.0117	.0176	.0247	.0328
	.01	.0002	.0008	.0020	.0041	.0071	.0112	.0164	.0225
11	.05	.0004	.0011	.0027	.0052	.0087	.0134	.0191	.0258
	.01	.0002	.0006	.0014	.0030	.0053	.0085	.0126	.0175
12	.05	.0002	.0008	.0019	.0038	.0066	.0103	.0150	.0205
	.01	.0001	.0004	.0010	.0022	.0040	.0065	.0098	.0138

TABLE A.19.   LOWER PERCENTAGE POINTS FOR WILKS'S LIKELIHOOD-RATIO U TEST (CDF)
(<u>cont.</u>)

ν_1	α	ν_2=28	30	40	60	80	100	120	140
					p=2				
7	.05	.4521	.4753	.5681	.6822	.7490	.7927	.8235	.8463
	.01	.3762	.4002	.4985	.6246	.7007	.7514	.7875	.8144
8	.05	.4197	.4430	.5374	.6561	.7268	.7736	.8067	.8314
	.01	.3473	.3710	.4695	.5987	.6782	.7317	.7700	.7988
9	.05	.3909	.4142	.5095	.6318	.7059	.7554	.7907	.8171
	.01	.3219	.3452	.4433	.5748	.6571	.7130	.7534	.7839
10	.05	.3652	.3882	.4839	.6090	.6861	.7380	.7753	.8033
	.01	.2993	.3222	.4195	.5525	.6372	.6954	.7376	.7696
11	.05	.3420	.3648	.4603	.5876	.6673	.7214	.7605	.7900
	.01	.2792	.3015	.3977	.5317	.6184	.6785	.7224	.7558
12	.05	.3211	.3435	.4386	.5675	.6493	.7054	.7462	.7771
	.01	.2611	.2829	.3778	.5123	.6006	.6624	.7078	.7426
					p=3				
7	.05	.3281	.3525	.4540	.5874	.6695	.7247	.7642	.7938
	.01	.2648	.2884	.3901	.5304	.6199	.6813	.7257	.7594
8	.05	.2943	.3180	.4188	.5552	.6411	.6996	.7418	.7737
	.01	.2358	.2584	.3578	.4993	.5917	.6559	.7029	.7388
9	.05	.2652	.2882	.3874	.5256	.6146	.6759	.7206	.7546
	.01	.2110	.2327	.3293	.4709	.5655	.6322	.6814	.7191
10	.05	.2399	.2621	.3593	.4983	.5897	.6535	.7004	.7362
	.01	.1899	.2105	.3040	.4448	.5411	.6098	.6610	.7004
11	.05	.2179	.2392	.3339	.4730	.5663	.6322	.6811	.7186
	.01	.1716	.1911	.2813	.4208	.5183	.5887	.6415	.6825
12	.05	.1985	.2189	.3110	.4495	.5442	.6120	.6625	.7016
	.01	.1556	.1741	.2610	.3987	.4969	.5687	.6230	.6654

TABLE A.19. LOWER PERCENTAGE POINTS FOR WILKS'S LIKELIHOOD-RATIO U TEST (CDF) (<u>cont</u>.)

ν_1	α	ν_2=28	30	40	60	80	100	120	140
					p=4				
7	.05	.2380	.2614	.3636	.5073	.6000	.6640	.7105	.7458
	.01	.1868	.2084	.3066	.4526	.5507	.6198	.6709	.7100
8	.05	.2061	.2282	.3270	.4711	.5670	.6342	.6836	.7214
	.01	.1604	.1805	.2739	.4183	.5184	.5902	.6437	.6851
9	.05	.1796	.2003	.2951	.4385	.5365	.6063	.6582	.6982
	.01	.1387	.1573	.2457	.3876	.4887	.5626	.6183	.6617
10	.05	.1573	.1767	.2672	.4088	.5082	.5801	.6341	.6760
	.01	.1206	.1378	.2213	.3598	.4615	.5368	.5943	.6394
11	.05	.1384	.1565	.2426	.3817	.4819	.5555	.6113	.6549
	.01	.1055	.1214	.1999	.3347	.4363	.5128	.5717	.6183
12	.05	.1224	.1392	.2209	.3570	.4574	.5323	.5897	.6348
	.01	.0928	.1074	.1812	.3119	.4129	.4902	.5503	.5982
					p=5				
7	.05	.1717	.1930	.2909	.4384	.5383	.6090	.6613	.7015
	.01	.1313	.1503	.2412	.3869	.4901	.5650	.6212	.6648
8	.05	.1435	.1630	.2550	.4001	.5020	.5755	.6306	.6733
	.01	.1086	.1257	.2098	.3511	.4550	.5320	.5906	.6364
9	.05	.1208	.1385	.2244	.3660	.4688	.5444	.6018	.6467
	.01	.0907	.1060	.1834	.3196	.4233	.5016	.5620	.6098
10	.05	.1023	.1184	.1983	.3355	.4384	.5155	.5748	.6215
	.01	.0762	.0899	.1611	.2916	.3944	.4736	.5354	.5847
11	.05	.0872	.1018	.1759	.3082	.4105	.4886	.5494	.5976
	.01	.0645	.0767	.1420	.2667	.3680	.4476	.5105	.5611
12	.05	.0748	.0879	.1565	.2836	.3848	.4635	.5254	.5750
	.01	.0550	.0659	.1257	.2444	.3439	.4234	.4871	.5387

TABLE A.19.   LOWER PERCENTAGE POINTS FOR WILKS'S LIKELIHOOD-RATIO U TEST (CDF)
              (cont.)

ν_1	α	ν_2=28	30	40	60	80	100	120	140
					p=6				
7	.05	.1228	.1416	.2322	.3788	.4831	.5590	.6159	.6601
	.01	.0916	.1078	.1894	.3309	.4366	.5155	.5758	.6231
8	.05	.0990	.1155	.1983	.3396	.4445	.5225	.5821	.6287
	.01	.0730	.0870	.1604	.2949	.3998	.4800	.5423	.5917
9	.05	.0804	.0950	.1701	.3054	.4097	.4891	.5506	.5993
	.01	.0588	.0709	.1366	.2637	.3669	.4477	.5114	.5625
10	.05	.0659	.0786	.1467	.2752	.3783	.4584	.5213	.5716
	.01	.0477	.0582	.1169	.2364	.3374	.4182	.4828	.5353
11	.05	.0543	.0655	.1270	.2486	.3497	.4300	.4940	.5456
	.01	.0391	.0481	.1006	.2126	.3107	.3911	.4563	.5097
12	.05	.0451	.0550	.1104	.2251	.3238	.4038	.4684	.5211
	.01	.0322	.0401	.0869	.1916	.2867	.3661	.4316	.4857
					p=7				
7	.05	.0870	.1031	.1847	.3270	.4336	.5131	.5738	.6214
	.01	.0633	.0767	.1483	.2829	.3891	.4706	.5340	.5843
8	.05	.0675	.0812	.1535	.2880	.3936	.4745	.5374	.5873
	.01	.0485	.0597	.1222	.2475	.3514	.4334	.4983	.5505
9	.05	.0529	.0645	.1284	.2545	.3581	.4395	.5039	.5556
	.01	.0376	.0470	.1013	.2174	.3182	.3998	.4657	.5193
10	.05	.0419	.0517	.1079	.2255	.3263	.4076	.4729	.5260
	.01	.0295	.0373	.0845	.1915	.2886	.3695	.4357	.4903
11	.05	.0334	.0417	.0912	.2003	.2978	.3784	.4442	.4983
	.01	.0233	.0299	.0709	.1693	.2624	.3419	.4081	.4633
12	.05	.0268	.0339	.0774	.1784	.2723	.3517	.4176	.4724
	.01	.0186	.0241	.0598	.1500	.2390	.3167	.3826	.4382

TABLE A.19. LOWER PERCENTAGE POINTS FOR WILKS'S LIKELIHOOD-RATIO U TEST (CDF)
(cont.)

ν_1	α	ν_2=170	200	240	320	440	600	800	1000
					p=2				
7	.05	.8713	.8893	.9068	.9291	.9478	.9614	.9709	.9766
	.01	.8441	.8656	.8865	.9135	.9362	.9527	.9643	.9713
8	.05	.8586	.8782	.8972	.9217	.9423	.9573	.9678	.9741
	.01	.8306	.8538	.8764	.9055	.9302	.9483	.9609	.9686
9	.05	.8463	.8674	.8880	.9145	.9369	.9533	.9647	.9716
	.01	.8177	.8424	.8665	.8978	.9244	.9439	.9576	.9659
10	.05	.8344	.8569	.8790	.9075	.9316	.9493	.9617	.9692
	.01	.8053	.8314	.8570	.8904	.9188	.9396	.9543	.9633
11	.05	.8228	.8468	.8702	.9006	.9264	.9454	.9587	.9668
	.01	.7932	.8207	.8478	.8831	.9132	.9355	.9511	.9607
12	.05	.8116	.8368	.8616	.8938	.9213	.9415	.9557	.9644
	.01	.7816	.8104	.8388	.8759	.9078	.9314	.9480	.9581
					p=3				
7	.05	.8265	.8503	.8735	.9034	.9287	.9471	.9601	.9679
	.01	.7969	.8243	.8511	.8859	.9156	.9373	.9526	.9619
8	.05	.8092	.8350	.8603	.8931	.9209	.9413	.9556	.9643
	.01	.7788	.8083	.8373	.8750	.9073	.9311	.9478	.9580
9	.05	.7925	.8203	.8475	.8830	.9133	.9356	.9512	.9608
	.01	.7616	.7930	.8239	.8645	.8993	.9250	.9432	.9542
10	.05	.7764	.8060	.8351	.8732	.9059	.9300	.9469	.9573
	.01	.7451	.7782	.8110	.8542	.8915	.9191	.9386	.9505
11	.05	.7609	.7922	.8231	.8637	.8986	.9245	.9427	.9539
	.01	.7292	.7640	.7985	.8442	.8838	.9132	.9341	.9469
12	.05	.7458	.7787	.8113	.8543	.8915	.9190	.9385	.9505
	.01	.7139	.7502	.7864	.8344	.8763	.9075	.9297	.9433

TABLE A.19.   LOWER PERCENTAGE POINTS FOR WILKS'S LIKELIHOOD-RATIO U TEST (CDF)
              (<u>cont</u>.)

ν_1	α	ν_2=170	200	240	320	440	600	800	1000
					p=4				
7	.05	.7852	.8141	.8424	.8792	.9105	.9335	.9497	.9595
	.01	.7540	.7864	.8184	.8604	.8963	.9228	.9415	.9529
8	.05	.7638	.7951	.8259	.8662	.9007	.9261	.9440	.9549
	.01	.7320	.7668	.8013	.8467	.8859	.9149	.9354	.9480
9	.05	.7433	.7768	.8100	.8535	.8910	.9188	.9384	.9504
	.01	.7111	.7480	.7848	.8335	.8758	.9072	.9295	.9432
10	.05	.7237	.7593	.7946	.8412	.8816	.9116	.9329	.9459
	.01	.6912	.7301	.7690	.8207	.8659	.8997	.9237	.9384
11	.05	.7049	.7423	.7796	.8292	.8724	.9046	.9274	.9415
	.01	.6721	.7128	.7536	.8083	.8563	.8923	.9180	.9338
12	.05	.6867	.7259	.7651	.8175	.8633	.8976	.9221	.9371
	.01	.6538	.6961	.7388	.7963	.8469	.8850	.9124	.9292
					p=5				
7	.05	.7466	.7800	.8129	.8561	.8931	.9204	.9397	.9514
	.01	.7143	.7512	.7878	.8361	.8779	.9089	.9308	.9443
8	.05	.7217	.7577	.7934	.8405	.8812	.9114	.9328	.9458
	.01	.6890	.7284	.7677	.8200	.8655	.8994	.9235	.9383
9	.05	.6979	.7363	.7746	.8255	.8697	.9026	.9260	.9403
	.01	.6649	.7066	.7484	.8044	.8534	.8902	.9164	.9325
10	.05	.6753	.7159	.7565	.8109	.8584	.8939	.9193	.9348
	.01	.6421	.6858	.7300	.7893	.8417	.8811	.9094	.9268
11	.05	.6536	.6962	.7391	.7966	.8473	.8854	.9127	.9295
	.01	.6205	.6660	.7122	.7747	.8303	.8723	.9025	.9212
12	.05	.6330	.6773	.7221	.7828	.8365	.8771	.9062	.9242
	.01	.5998	.6469	.6950	.7606	.8191	.8636	.8958	.9156

TABLE A.19.   LOWER PERCENTAGE POINTS FOR WILKS'S LIKELIHOOD-RATIO U TEST (CDF)
(cont.)

ν_1	α	$\nu_2=170$	200	240	320	440	600	800	1000
					p=6				
7	.05	.7104	.7478	.7849	.8339	.8762	.9077	.9299	.9435
	.01	.6774	.7181	.7589	.8131	.8603	.8955	.9206	.9359
8	.05	.6823	.7224	.7626	.8160	.8625	.8972	.9218	.9369
	.01	.6490	.6924	.7361	.7946	.8459	.8845	.9120	.9290
9	.05	.6557	.6983	.7412	.7987	.8491	.8869	.9139	.9305
	.01	.6223	.6680	.7143	.7768	.8320	.8738	.9037	.9222
10	.05	.6305	.6753	.7207	.7819	.8360	.8768	.9060	.9241
	.01	.5972	.6448	.6935	.7596	.8186	.8633	.8956	.9155
11	.05	.6065	.6533	.7010	.7657	.8232	.8669	.8984	.9178
	.01	.5734	.6228	.6735	.7430	.8054	.8531	.8876	.9090
12	.05	.5837	.6322	.6819	.7499	.8108	.8572	.8908	.9116
	.01	.5508	.6017	.6544	.7270	.7926	.8431	.8797	.9025
					p=7				
7	.05	.6760	.7171	.7580	.8125	.8599	.8952	.9204	.9358
	.01	.6427	.6869	.7313	.7909	.8432	.8825	.9105	.9278
8	.05	.6452	.6891	.7332	.7924	.8443	.8833	.9111	.9282
	.01	.6117	.6586	.7061	.7703	.8271	.8700	.9009	.9199
9	.05	.6162	.6625	.7095	.7730	.8291	.8716	.9020	.9208
	.01	.5828	.6319	.6820	.7504	.8115	.8579	.8914	.9121
10	.05	.5888	.6373	.6868	.7542	.8144	.8602	.8931	.9135
	.01	.5557	.6066	.6591	.7313	.7963	.8461	.8822	.9045
11	.05	.5630	.6133	.6650	.7361	.8000	.8490	.8844	.9063
	.01	.5302	.5827	.6373	.7129	.7816	.8346	.8731	.8971
12	.05	.5385	.5904	.6442	.7186	.7861	.8380	.8758	.8993
	.01	.5061	.5600	.6164	.6952	.7673	.8233	.8642	.8897

Source: Extracted by permission from tables produced by F. J. Wall, 1968
(290 Alamosa Rd. NW, Albuquerque, N.Mex.) with assistance from the Dikewood
Corp. of Albuquerque.

Glossary of Symbols

Some symbols in the text have been used in several ways, chiefly because of well-entrenched traditions. However some traditions have been overthrown to minimize double usage or to achieve better conformity with the general notational conventions set out below:

1. Greek letters denote probabilities, parameters, and fixed treatment effects from a population or distribution. (Major exceptions: χ^2 for percentage points of the chi-square distribution; Σ and Π for the operations of summation and multiplication.)

2. Capital italic letters denote random variables (including estimators, random effects of classification factors, and random experimental or sampling errors), factors investigated, and a few other quantities of transient interest. (Major exceptions: $P[\ldots]$, $E[\ldots]$, $V[\ldots]$, and $\text{Cov}[\ldots]$ for probability and parametric expectations, variances, and covariances; T^2 and U for multivariate sample statistics.)

3. Lowercase italic letters denote fixed levels or sample values taken by random variables, estimates of parameters of populations, the number of levels or classes of factors, the number of replications or items in a set, and subscripts for identification of individual symbols in a set.

4. Boldface roman letters denote matrices (capitals) or vectors (lowercase letters). The term "vector" refers to a matrix array of symbols or numbers having a single row or column (arrays of population parameters are denoted by Greek letters as usual, but estimators and estimates are not distinguished by case).

5. A circumflex (^) above a Greek symbol denotes an estimator or estimate of a population parameter; over a capital italic letter, it denotes a predicted value of a random variable for a given set of conditions.

6. Subscripts replaced by dots indicate summation over the range of the subscripts replaced.

7. A bar above a symbol denotes the arithmetic mean (exception: $\bar{H}$ indicates an alternative hypothesis).

The most pertinent sections of the text are indicated for each symbol or definition. In most cases the first section referenced provides the most fundamental or complete description.

B.1 GREEK LETTERS

α (alpha): Probability of a random variable exceeding (or exceptionally, falling short of) a specified point of its density or sampling distribution; most often, probability of making a Type I error, i.e., rejecting a true hypothesis [1.4, 1.5, 1.6].

α_i: Fixed effect of ith level of factor A [2.4, 3.7, 4.3, 4.4, 5.8, 7.6, 8.2, 8.3, 8.4].

β (beta): Probability of Type II error, i.e., accepting a false hypothesis [1.6, 2.1.3, 2.2.1.3].

β_j: Parameter denoting slope of linear regression of random variable Y on fixed variable x_j (change in Y per unit increase in x_j) [1.8, 2.2.2, 3.2, 4.2]; fixed effect of jth level of factor B [2.4, 3.7, 4.3, 4.4, 5.8, 7.6, 8.2, 8.3, 8.4].

$\beta_{jj'}$: Parameter denoting regression of random variable Y on $x_j x_{j'}$, i.e., a second degree polynomial ($j = j'$), or interaction of two quantitative factors ($j \neq j'$) [4.2, 9.1, 9.2, 9.3].

β_0: Parameter denoting origin of regression of random variable Y on fixed variables), i.e., value of Y expected for zero value(s) of fixed variable(s) [1.8, 2.2.2, 4.2, 9.1].

γ (gamma): Probability that a confidence interval will be shorter than a specified number of standard deviations [1.5.4.2]; probability that "slippage" in coefficient of multiple determination will be less than a specified amount [4.2.7].

γ_j: Fixed effect of jth column of Latin square [7.2].

γ_k: Fixed effect of kth level of factor C [2.4, 8.2.2].

$(\alpha\beta)_{ij}$, $(\alpha\gamma)_{ik}$, $(\beta\gamma)_{jk}$, $(\alpha\beta\gamma)_{ijk}$: Fixed effects of interactions among factors A, B, and C [2.4, 3.7, 4.3, 4.4, 5.8, 7.6, 8.2, 8.3, 8.4].

Δ (capital delta): Parametric (population) mean difference between two groups or two sides of a contrast [2.1.3, 2.2.1, 2.4.6, 5.7, 5.9, 6.3]; loss in response expected for dose smaller than optimal [2.2.2.2].

Δ_d: Specified difference one wishes to detect [1.7.3, 2.2.1.3, 2.3.4, 2.4.3.6, 5.5, 7.3, 8.3].

δ_j (delta): Fixed effect of jth block [5.3.1, 5.8, 6.2, 6.3, 9.4].

ε (epsilon): Arbitrarily small value [1.5.2, 4.2.7]; parameter of nonadditivity [5.3.2].

ε_i: Probability of ith outcome for discrete random variable [1.3.5, 1.3.6, 1.5.3].

η (eta): Number of events for binomial or multinomial distribution or for contingency table of frequency data [1.4.1, 1.4.2, 1.7, 1.8.1.3]. (Note: The symbol n has been widely used for this purpose, but the number of events is a parameter of a specific distribution; hence a Greek symbol is used.)

θ (theta): Mean of exponential distribution [1.3.5]; arbitrary or unspecified parameter [1.5]; fixed effect associated with one of several Latin squares [7.2, 8.1.3, 8.2.3].

λ (lambda): Only parameter of Poisson distribution [1.4.3, 1.5.3, 1.5.4.4, 1.6.7]; number of comparisons of any two treatments in same block of balanced incomplete block design [6.2.1, 8.1.6]; generalized exponent on defining relation used to create incomplete blocks for complex factorial experiments [6.5].

λ_i: Number of comparisons in same block of two treatments in partially balanced incomplete block design [6.2.2]; ith characteristic (latent) root or eigenvalue of matrix [4.1.2, 9.1.2]; moment parameters of rotatable response surface designs [9.3.2].

μ (mu): Mean of distribution or population [1.3, 1.4, 1.5, 1.6.2, 2.1].

μ_D: Mean of distribution of differences between two random variables [1.6.6.6].

μ_d: Magnitude of mean one wishes to detect [1.6].

μ_i: The ith central moment of distribution [1.3.6]; mean of ith population [1.4, 1.5.4.2, 1.6, 4.5, 8.4].

μ_0: Hypothesized value for population mean [1.6, 4.5].

$(\mu_Y|x)$: Mean of distribution of random variable Y for given value x of related variable [1.8.2, 4.2.5].

ν (nu): Number of degrees of freedom associated with distribution of sample statistic [1.4.6].

ξ_i (xi): Orthogonal polynomial coefficients [2.2.2, 2.4.3.7, 6.4, 8.2.2, A.7].

Π (capital pi): Operational symbol indicating product of values following it [1.5.3].

π (pi): Ratio of circumference to diameter of circle, 3.1416... [1.4, 1.5, 1.8].

ρ (rho), ρ_{12}, ρ_{xY}, etc.: Correlation between two variables in population [1.3.6, 1.6.6.6, 1.8, 2.2.1, 3.5, 8.2, 8.4].

ρ_i: Fixed row total in contingency table of frequency date [1.7.3]; fixed effect of ith row of Latin square [7.2]; fixed effect of ith period in changeover design [8.1].

ρ_0: Hypothesized value for population correlation between two variables [1.8.1].

Σ (capital sigma): Operational symbol indicating summation of values following it [1.3.3].

σ (sigma): Standard deviation (square root of variance) of distribution or population [1.3.4.2, 1.3.6, 1.4, 1.5, 1.6, 1.8, 2.1.3].

σ_D: Standard deviation of population of differences [1.6.6.6].

σ_d: Magnitude of standard deviation one wishes to detect [1.6.4, 1.6.5].

σ_{ij}: First central product moment (covariance) of joint distribution of two variables [1.3.6, 1.6.6.6, 1.8, 2.1.4, 4.5].

σ^2: Variance of population, distribution of random variable, or random experimental errors associated with variable [1.3.4.2].

$\sigma_A^2, \sigma_B^2, \sigma_C^2$, etc.: Components of variance associated with random classification factors A, B, C, etc. [2.4].

σ_D^2: Component of variance among determinations within samples on same subjects [2.3.6]; component of variance among random blocks [6.2, 8.1].

σ_i^2: Variance of ith population [1.4.7, 1.6.5, 1.6.6, 2.1, 2.3, 4.5, 5.3.3, 5.6, 8.2]; component of variance associated with ith random classification factor [2.3, 2.4.2, 4.4.5].

σ_0^2: Hypothesized value for population variance [1.6.4].

σ_T^2: Component of variance among random groups [2.1, 2.3].

σ_U^2: Component of variance among random samples within subjects [2.3].

$(\sigma_Y^2|x)$: Variance of random variable Y within fixed levels of variable x [1.8.2, 3.2, 4.2].

   τ_i (tau): Fixed effect of ith treatment (or classification) group or ith treatment combination [2.1].

   τ_d: Magnitude of treatment effect one wishes to detect [2.1.3, 5.5, 7.3].

   ϕ (phi): Relative difference(s) to be detected (abscissa scale) on power charts for analysis of variance or orthogonal contrasts [2.1.3, 2.2.1.3, 2.3.4, 2.4.3.6, 4.5, 5.9, 7.3, 8.3]; degree of resolution of fractionally replicated designs [6.1, 6.3.4].

   $\chi_{\alpha,\nu}^2$ (chi-square): Percentage point of chi-square distribution with ν df, exceeded with probability α [1.4.6, 1.5.4, 1.6, 1.7, 2.1.5.5, 2.3, 4.5.2, 6.2.1.3, 8.4, A.3]. (Note: The long tradition of using a Greek symbol for this quantity constitutes an exception to the general conventions of notation established.)

   $\chi_{B,\alpha,m,\nu}^2$: Percentage point of Bonferroni chi-square for m correlated tests of subcontingency tables with ν df [1.7.3, A.11].

   Ψ (capital psi): Constant that maximizes or minimizes a function [1.5.2].

   ω (omega): Probability of occurrence of one of two events of binomial distribution [1.4.1, 1.7]. (Note: The symbol p has been widely used for this purpose, but the probability is a parameter of the distribution; hence a Greek symbol is used.)

   ω^*: Probability of new treatment being preferred over old treatment in all-or-none sequences of paired comparisons [6.3.4.7].

   ω_i: Probability of occurrence of ith of several events of multinomial distribution [1.4.2, 1.7]; probability of a binomial event in ith population [6.3.4.7].

   ω_0: Hypothesized value of probability for binomial or multinomial event [1.7].

B.2   CAPITAL ITALIC LETTERS

   A, B, C, etc.: Treatment or classification factors studied simultaneously in an experiment [2.4, 3.7, 4.4, 5.8, 6.3, 6.4, 6.5, 7.6, 7.8.4, 8.2, 8.3]; levels or classes of treatment factor in Latin square design (traditional exception to conventional notation) [7.1, 8.1].

   A_i, B_i, C_k, etc.: Random effects of classification factors A, B, C, etc. [2.4].

   AB, AC, BC, ABC: Types of interactions among factors A, B, and C [6.3, 6.4, 7.8.4].

   $(AB)_{ij}$, $(AC)_{ik}$, $(BC)_{jk}$, $(ABC)_{ijk}$: Random effects of interactions among classification factors A, B, and C [2.4.4].

(AB), (AB^2), (AB^2C^2), etc.: Orthogonal components of interactions in factorial experiments with 3 or more levels per factor in incomplete block designs [6.2.3.2, 6.4, 6.5].

C_p: Statistic for comparison of p variable subsets of k candidate variables in multiple regression analysis [4.2.6].

$Cov[\ldots, \ldots]$: Population covariance of two random variables bracketed [1.3.6, 1.8.1, 4.2, 6.2]. (Note: This traditional notation is an exception to the rule requiring Greek symbols for population constants.)

D: Fourth factor in experiment with 4 or more factors [2.4.4, 6.3, 6.4]; generalized distance statistic [4.5].

$D_j, D_{(i)j}$: Random effect of jth block [5.7, 8.1.2, 8.2].

$D_{(ijk)l}$: Random effect of one laboratory determination on sample of material [2.3.6].

E: Relative efficiency of design or sampling plan [2.3.4, 3.5, 4.3, 5.6, 6.2, 6.3, 7.4, 8.1, 8.2, 8.3]; fifth factor in experiment with 5 or more factors [6.3, 6.4].

E_i, E_{ij}, E_{ijk}: Expected number of observations in category of frequency data [1.7]. (Note: This traditional notation is an exception to the rule requiring Greek symbols for parametric values.)

E_i, $E_{(i)j}$, $E_{(ij)k}$, etc.: Random experimental error associated with particular experimental unit whose response to treatment is specified by a model [1.8.2, 2.1].

$E[\ldots]$: Mathematical expected value (population average) of random variable bracketed [1.3.6, 1.4, 1.5, 1.7, 1.8, 2.1, 2.2, 2.3, 2.4, 3.4, 4.2.3, 4.4, 6.2, 8.4]. (Note: This traditional notation is an exception to the rule requiring Greek symbols for population constants.)

F: Fisher's variance-ratio variable [1.4.8, 1.6.5, 2.1.3, 2.2.1, A.5]; sixth factor in experiment with 6 or more factors [6.3, 6.4].

G: Arbitrary random variable of transient interest; seventh factor in experiment with 7 or more factors [6.3, 6.4].

H: Hypothesis tested (commonly termed null hypothesis) [1.6]; eighth factor in experiment with 8 or more factors [6.3, 6.4].

$\bar{H}$: Alternative to hypothesis tested [1.6].

I: Fundamental identity of fractionally replicated design, denotes inestimable effects [6.3.4, 6.4.3, 6.5].

J: Randomization or restriction error in models for blocked designs [5.3, 5.8].

L: Likelihood of occurrence of sample of observations [1.5.3]; linear function of mean squares [2.3, 2.4.4].

MS: Mean square, i.e., linear function of 2 or more quadratic forms of random variable [2.1, 2.2, 2.3, 2.4.4].

N: Abbreviation for normal distribution [1.4.5, 1.8.2, 2.1, 2.4, 3.2].

O_i, O_{ij}, O_{ijk}: Observed number of units for category of frequency data [1.7]. (Note: This traditional notation is an exception to the rule requiring lowercase letters for sample values.)

$P[\ldots]$: Probability of occurrence of event bracketed [1.3.5, 1.4, 1.5, 2.2.1, 4.2.7]. (Note: This traditional notation is an exception to the rule requiring Greek symbols for probabilities.)

Q: Random variable distributed exactly or approximately as chi-square [1.4, 1.5, 1.6.8, 2.1.2].

Q_k: Linear function of totals (or means) of sampling variables associated with contrasts [2.2.1].

R: Residues of modular arithmetic operations, used to allocate treatment combinations to incomplete blocks [6.3, 6.4, 7.8.4]; coefficient of multiple correlation among 3 or more variables [4.2].

R^2: Squared multiple correlation or "coefficient of multiple determination," fraction of variation in dependent variable attributable to variation in independent variables [4.2]. (Note: Traditionally, this notation is used for sample values, as well as for the random variable itself.)

R_j: Random effect of jth replicate or repetition of set of treatments [5.9, 6.3, 6.4, 8.3].

S: Standard deviation for samples drawn from population; estimator of σ [1.3.4.2, 1.3.6.2, 1.4.4, 1.4.7, 1.5.2, 1.5.4.2, 1.6.3].

S^2: Variance for samples drawn from population; estimator of σ^2 [1.3.4.2, 1.3.6.2, 1.5.2, 1.5.4.3, 1.6.4, 2.1.2.3].

S_i^2: Variance of samples drawn from ith population [1.6.5, 2.1.5.5].

$(S_Y^2|x)$: Variance of samples of Y within fixed levels of x [1.8.2].

S_{ij}: Covariance of two random variables for samples drawn from bivariate population; estimator of σ_{ij} [1.3.6, 4.3, 4.5].

SS: Sum of squared deviations of sample variables from sample mean variable [2.1.2.2, 2.1.3, 4.2.2].

T: Variable following Student's t distribution [1.4, A.4].

T^2: Hotelling's multivariate distance statistic, analogous to the univariate Student's t [4.5, 4.6, 8.4, A.18]. (Note: This traditional notation is an exception to the rule requiring lowercase letters for sample statistics.)

T_α^2: Percentage point of multivariate T^2 distribution, exceeded with probability α [4.5, 4.6, 8.4, A.18].

U: Wilks's multivariate likelihood-ratio statistic [4.6, 8.4, A.19]. (Note: This notation is an exception to the rule requiring lowercase letters for sample statistics.)

U_α: Lower percentage point of multivariate U distribution, exceeded with probability $1-\alpha$ [4.6, 8.4, A.19].

$U_{(ij)k}$: Random error among samples taken from same subject [2.3].

V_i: Variance of treatment mean for ith sampling plan [2.3.4].

$V[\ldots]$: Population variance of random variable bracketed [1.3.6, 1.4.4, 1.4.5, 1.7, 1.8, 2.1]. (Note: This traditional notation is an exception to the rule requiring Greek symbols for population parameters.)

Y: Dependent random variable of primary interest to experimenter [1.3].

Y_D: Dependent random variable describing differences [1.6.6.6].

Y_m: Predicted maximum (or minimum) response for optimal dose of treatment [2.2.2.2].

Y_s: Predicted response at stationary point of response surface [9.1].

Z: Standard normal variable with zero mean and variance equal to unity [1.4, 1.5.4, 1.6.2, 1.6.6, 1.8.1, A.2].

B.3 LOWERCASE ITALIC LETTERS

a, b, c, etc.: Number of levels or classes of factors A, B, C, etc. [2.4, 3.7, 5.8, 7.6, 8.2, 8.3]; higher of two levels of each of factors A, B, C, etc., in symmetrical designs [6.3].

a_i: Coefficient of level of ith factor, used in allocating treatment combinations of factors with 3 or more levels to incomplete blocks [6.4, 6.5].

a_{ii}: Second order, off-diagonal elements of the $X'X$ matrix for nonorthogonal, central composite, response surface designs [9.3.1].

a_{ij}, b_{ij}, c_{ij}, etc.: Elements in ith row and jth column of matrices A, B, C, etc. [4.1, 4.2, 4.3, 4.4].

$a_{i,n}$: Coefficients of residual differences in test of normality [2.1.5.3, A.13.1].

b: Number of blocks in incomplete block design [6.2, 6.3] or in response surface block design [9.4].

b_j: Sample estimate of slope of linear regression of random variable Y on fixed variable x_j [1.8.2, 2.2.2, 3.4, 4.2].

b_0: Sample estimate of origin of regression of Y on one or more fixed variables [1.8.2, 2.2.2, 4.2].

c: Arbitrary constant [1.3, 1.5.4.2, 2.1.5.3]; number of columns in arbitrary contingency table [1.7.3] or matrix [4.1].

c_i: Arbitrary constants; coefficients of treatment totals (or means) or other quantities in linear functions and contrasts [2.2, 2.3.4, 2.4, 3.4, 4.6, 5.3.3, 5.7, 6.2, 6.3, 7.3, 8.3, 8.4]; costs of experimental units and subsamples [2.3.4]; design constants for partially balanced incomplete blocks [6.2.2].

d: Width of class interval [1.3.4.4]; number of determinations per laboratory sample [2.3.6]; number of dependencies in matrix of nonfull rank [4.1]; univariate distance statistic [4.5].

d_{ij}: Difference for ith variable measured on jth pair of experimental units [4.5].

e: $e = 1 + \sum_{n=1}^{\infty} (1/n!) \simeq 2.71828$, where $n! = n(n-1)(n-2)...(1)$. It is the base number of natural (Napierien or hyperbolic) logarithms [1.3.5, 1.4, 1.5, 1.6.8, 1.8, 4.5].

e_i, $e_{(i)j}$, $e_{(ij)k}$, etc.: Residual (experimental) error associated with particular observation in sample [1.8, 2.1.5, 4.2.8, 5.3.3, 7.2.2].

f: Sample value of random variable following Fisher's variance-ratio distribution (traditionally, F) [1.4.8, 1.6.5, 1.8.2, 2.1]; number of factors in interaction [2.4.6, 6.4, 6.5].

f_{α,ν_1,ν_2}: Percentage point of F distribution with ν_1 and ν_2 df for numerator and denominator, exceeded with probability α [1.4.8, 1.6.5, 1.8.2, 2.1, A.5].

f_{ij}: Cofactor of element in ith row and jth column of square matrix [4.1.2].

f_{max} : Sample value of random variable that follows Hartley's f_{max} distribution [2.1.5.5, 8.2, A.6].

$f_{max,\alpha,t,\nu}$: Percentage point of F_{max} distribution (with t independent variances based on ν df each), exceeded with probability α [2.1.5.5, 8.2, A.6].

$f(\ldots)$, $g(\ldots)$, etc.: Functions of random variables bracketed [1.3.5, 1.4, 1.5, 1.6, 1.8].

g, h: Quantities of transient interest.

i, j, k, etc.: Subscripts used to index sets of data, estimates, parameters, or variables [1.3.3].

k: Number of classes in sample of grouped data [1.3.4.4]; number of experimental units in incomplete block [6.2, 6.3, 7.8.2, 8.1.6]; order of finite field [6.5.1.2]; number of independent variables in multiple regression [4.2, 9.2, 9.3, 9.4].

k_i: Coefficients of components of variance in expected mean squares for unbalanced data [2.3, 4.4.5].

m: Integer denoting multiple, power, or some other constant; arbitrary number of items in set; number of comparisons among means in set [2.2]; number of association classes in partially balanced incomplete block designs [6.2.2]; number of Latin squares in experiment [7.2, 8.1]; factor levels for axial points in central composite, response surface designs [9.3, 9.4].

m_i: The ith sample moment [1.5.3]; number of cells missing for ith treatment in randomized block design [5.4].

m_{ij} : Minor of element in ith row and jth column of square matrix [4.1.2]; element in ith row and jth column of orthogonal matrix [9.1.2].

m_j : Multipliers to make integer orthogonal polynomial coefficients [2.2.2.2, A.7].

ms: Sample mean square (sum of squares divided by degrees of freedom) [2.1].

n: Total number of observations on variable in experiment or sample [1.3, 1.4, 1.5, 1.6, 1.8, 2.1]; number of factors in symmetrical factorial experiment [2.4.6, 6.3, 6.4, 6.5.1]; number of subjects in multivariate analysis of repeated measurements [8.4.2].

p: Number of criteria (factors) forming dimensions of contingency table for frequency data [1.7.3.2]; number of levels per factor in symmetrical experiments [6.3, 6.4, 6.5.1]; number of parameters in multiple regression or covariance or in other linear models [4.2, 4.3, 4.4]; number of variables in multivariate analysis [4.5, 4.6]; number of periods of repeated measurement [8.4].

q: Sample value of random variable distributed exactly or approximately as chi-square [1.4.6, 1.6, 1.7, 1.8.1.3, 2.1.5.5, 4.5.2, 6.2.1.3, 8.4, A.3].

$q_{\alpha,t,\nu}$: Percentage point of distribution of Studentized range (with t items and ν df), exceeded with probability α [2.2.4, A.8].

q_k: Linear function of sample totals (or means) for contrasts [2.2, 2.4, 3.4, 5.3.3, 6.2, 6.3, 6.4, 6.5, 8.1.4, 8.2.2].

r: Number of replications (experimental units or subjects) per treatment or treatment combination (r_i for ith group if replication is unbalanced); number of blocks in most complete block designs [1.3.3, 1.4, 1.5.4.2, 1.6.5, 1.6.6, 1.8.2.5, 2.1]. (Note: The symbol n (or n_i) has been widely used to indicate amount of replication.) Number of rows in an arbitrary contingency table [1.7.3] or matrix [4.1].

r_0: Weighted average of unbalanced replications [2.1].

r_{12}, r_{xy}: Sample correlation between two variables (used without subscripts only when context is unambiguous) [1.3.4.3, 1.6.6.6, 1.8.1, 1.8.2, 3.5, 8.2].

$\bar{r}_h$: Harmonic mean of subclass replications in unbalanced factorial data [2.4.7].

$r[...]$: Rank of matrix bracketed [4.1.2, 4.4.3].

s: Sample standard deviation [1.3.4.2, 1.4.7, 1.5.4.2, 1.6.3, 1.6.6]; number of defining relations used to divide replicate into set of incomplete blocks or fractions [6.3, 6.4, 6.5].

s_b: Standard error of regression coefficient [1.8.2, 4.2.4].

s_D: Sample standard deviation of differences [1.6.6.6, 1.8.1, 8.2].

s_{ij}: Sample covariance for two variables [1.3.4.3, 1.8, 4.5, 4.6, 8.2, 8.4].

s^2: Sample variance [1.3.4.2, 1.3.6.2, 1.4, 1.5.4.3, 1.6.4, 1.6.6, 1.8.1].

s_i^2: Sample variance of ith group [1.4.8, 1.6.5, 1.6.6.2, 2.1, 2.3, 4.5, 5.3.3, 8.4].

$(s_y^2|x)$: Sample variance of y within levels of x [1.8.2].

sp: Sample sum of products of deviations of two variables from their sample means [1.8, 3.3, 4.3, 4.5, 4.6, 9.1].

ss: Sample sum of squares of deviations of variable from its sample mean (subscripted to indicate component parts of total sum of squares) [1.5.4.2, 1.8, 2.1].

$(ss_y|x)$: Sample sum of squared deviations for y within levels of x [1.8.2].

t: Sample value of random variable that follows Student's t distribution (any linear function of normally distributed variables divided by square root of sample variance of linear function) [1.4.7, 1.6.3, 1.6.6, 1.8, 2.2.1.2, 2.4.3.7, 3.4, 4.2, 4.3, 4.5, 6.2, 8.1.5, 8.2.2, 9.1, A.4]; number of treatments or treatment combinations [1.3.3, 1.8.2.5, 2.1].

$t_{\alpha,\nu}$: Percentage point of Student's t distribution (with ν df), exceeded with probability α [1.4.7, 1.4.8, 1.5.4.2, 1.6.3, 1.6.6, 1.8, 2.2, A.4].

t_B: Sample value of t statistic to be evaluated by use of Bonferroni t distribution for correlated comparisons [2.2.3.1, A.10].

$t_{B,\alpha,m,\nu}$: Percentage point of Bonferroni t distribution (with ν df), exceeded with probability α for set of m correlated comparisons [2.2.3.1, A.10].

t_D: Sample value of t statistic to be evaluated by use of Dunnett's t distribution for comparisons with control [2.2.5, A.9].

$t_{D,\alpha,m,\nu}$: Percentage point of Dunnett's t distribution (with ν df), exceeded with probability α for set of m comparisons with control [2.2.5, A.9].

u: Number of random observations per experimental unit or subject in nested classification (u_{ij} for unbalanced data) [2.3]; number of sets of aliases of factorial effects used to partition fractional replicate into blocks [6.3.4.3, 6.4.3.3, 6.5.1].

u_i: Element in finite field [6.5.1.2].

w: Sample value of Shapiro-Wilk statistic used for testing normality [2.1.5.3, A.13].

$w_{\alpha,n}$: Critical value of Shapiro-Wilk test for normality of sample of n observations, exceeded with probability 1-α [2.1.5.3, A.13.2].

w_i: Coefficients in weighted linear function [1.8.2.9, 2.4.7]; canonical variables, linear functions of fixed independent variables that produce a response surface [9.1.2].

x: Controllable, fixed variable presumed to be related to dependent (primary) random variable Y [1.3.4.3, 1.8.2, 2.2.2, 3.2, 4.2, 4.3, 4.4, 5.4, 6.3, 6.4, 6.5, 7.8.4, 9.1, 9.2, 9.3, 9.4].

x_e: Economic dose of treatment [2.2.2.2].

x_s : Level of factor that maximizes (or minimizes) response [2.2.2.2, 9.1].

y: Sample value of random variable Y of primary interest to the experimenter [1.3].

y_d: Sample difference between two observations of pair [1.6.6.6, 1.8.1.3].

y_0: Observed value of instrumental variable for unknown laboratory specimen, to be used in inverse prediction [1.8.2.8].

z: Sample value of random variable that follows standard normal distribution [1.4.4, 1.4.5, 1.6, 1.7, 1.8.1, 8.2, A.2].

$z_{1-\alpha}$: Cumulative percentage point of standard normal distribution, exceeded with probability α [1.4.4, 1.4.5, 1.5.4, 1.6, 1.7, 1.8.1, 8.2, A.2].

z_{12}: Transformed (normalized) sample correlation [1.8.1].

z_m : Block indicator variable in models for response surface block designs [9.4].

z_0 : Transformed hypothesized value of correlation [1.8.1].

B.4 BOLDFACE CAPITAL LETTERS (Matrices)

The determinant, transpose, cofactor, adjugate, inverse, generalized inverse, and rank of a square matrix A are denoted, respectively, by $|A|$, A', cof A, adj A, A^{-1}, $\bar{A}$, and $r(A)$. (See Sec. 4.1.)

A: Matrix of coefficients from arbitrary set of simultaneous equations [4.1.2].

A, B, C: Arbitrary matrices used to illustrate matrix operations [4.1]; submatrices of (X'X) [4.4].

B (capital beta): Matrix of second order polynomial parameters pertaining to response surface [9.1.2].

C: Inverse of matrix (X'X) [4.2]; matrix of variances and covariances of estimates of parameters [4.3, 4.4]; contrast matrix for testing means of periods of repeated measurement [8.4].

D: Arbitrary diagonal matrix; variance-covariance matrix of random effects in mixed model for unbalanced data [4.4.5.2].

E: Matrix of error sums of squares and products in analysis of multiple

covariance [4.3]; variance-covariance matrix of random errors in mixed model
for unbalanced data [4.4.5.2].

G: Generalized inverse of (X'X) [4.4].

G_{ij}: Submatrix of G [4.4].

I: Identity matrix, i.e., diagonal elements equal to unity and off-
diagonal elements equal to zero [4.1, 4.4, 8.4, 9.1.2].

K: Penrose generalized inverse matrix [4.1]; matrix of coefficients of
components of variance in expected mean squares for random model with unbalanced data [4.4.5.1].

M (capital MU): Matrix of population means with t rows (number of treatment combinations) and p columns (number of periods of repeated measurement
[8.4]; orthogonal matrix, i.e., one for which M'M = I [9.1.2].

O: Matrix of zeros [8.4].

P: Submatrix on diagonal of $(X'X)^{-1}$ [4.4]; matrix containing from 1 to
$p-1$ columns of p contrast coefficients of means for p periods of repeated measurement [8.4].

Q: Submatrix off diagonal of $(X'X)^{-1}$ [4.4].

R: Submatrix on diagonal of $(X'X)^{-1}$ [4.4.2]; diagonal matrix of ratios
of factorial components of variance to error component [4.4.5.2].

S: Matrix of sample variances and covariances among two or more related
groups or variables [4.5, 8.4].

S_D: Matrix of sample variances and covariances of paired differences for
p variables [4.5].

S_E: Error matrix for testing hypotheses in multivariate analysis of
means [4.6, 8.4]

S_H: Matrix for hypothesis tested in multivariate analysis of means [4.6,
8.4].

S_0: Matrix with equal diagonal elements (average of sample variances)
and equal off-diagonal elements (average of sample covariances) [8.4].

S_p: Matrix of pooled estimates of two or more variances and one or more
covariances [4.5, 8.4].

S_T (or S_A, S_B, S_{AB}, etc.): Matrix of totals of sums of squares and products (among covariates) for error plus treatment effect (or one of several
factorial effects) in analysis of multiple covariance for balanced data [4.3].

T: Matrix containing from 1 to $t-1$ columns of t contrast coefficients of
t treatment means [8.4].

V: Matrix of variances and covariances of observations on dependent
variable Y in generalized analysis of unbalanced data [4.4.5].

X: Matrix of n rows (number of experimental units or subjects) and p
columns (number of parameters to be estimated), where each element is value of
independent variable or indicator (0,1) of structure of classification for
factorial data [4.2, 4.4, 9.1, 9.2, 9.3].

Y: Matrix of n rows (number of subjects) and p columns (number of periods of repeated measurement) of observations on random variable Y [8.4].

Λ (capital lambda): Diagonal matrix of characteristic (latent) roots or
eigenvalues of matrix [9.1.2].

Σ (capital sigma): Matrix of population variances and covariances among
two or more variables or periods of repeated measurement [4.5, 8.4].

Σ_D: Matrix of population variances and covariances of paired differences
for p variables [4.5].

Σ_0: Homogeneous variance-covariance matrix for p periods of repeated measurement of population, with diagonal elements σ^2 and off-diagonal elements $\rho\sigma^2$ [8.4].

B.5 BOLDFACE LOWERCASE LETTERS (Vectors)

A column vector a after being transposed to a row vector is denoted by a'.

a: Arbitrary column vector containing r elements (rows) [4.1].

c: Column vector of constants describing linear functions of means for which one desires simultaneous confidence intervals in multivariate analysis [4.5, 8.4].

d: Column vector of numbers of degrees of freedom associated with two or more random factorial effects in unbalanced data [4.4.5]; column vector of sample mean differences [4.5].

e: Column vector of n sampling errors (residuals) relative to model for dependent variable Y [4.2, 4.4, 9.1, 9.2]; column vector of p sums of products of covariates with dependent variable [4.3].

m_i: The ith column of an orthogonal matrix M [9.1.2].

0: Vector of zeros [9.1.2].

s: Column vector of analogous sums of squares, used to estimate variance components in unbalanced factorial data [4.4.5.1].

s_T (or s_A, s_B, s_{AB}, etc.): Column vector of totals of sums of products (of covariates with dependent variable) for error plus treatment effect (or one of several factorial effects) in analysis of multiple covariance for balanced data [4.3].

w: Column vector of canonical variables, linear functions of fixed independent variables that produce response surface [9.1.2].

x: Column vector of factor levels in response surface design [9.1].

x_s: Column vector whose elements provide the coordinates of stationary point of response surface [9.1.2].

y: Column vector of n observations on dependent variable Y for n experimental units or subjects [4.1.3, 4.2, 4.4, 9.1, 9.2]; column vector of p repeated measurements on subject [8.4].

$\bar{y}$: Column vector of sample means for p related variables [4.5, 8.4].

z: Deviations $x-x_s$ of design factor levels from optimal levels [9.1.2].

β (beta): Column vector of p parameters, each a partial regression coefficient or effect of treatment or factorial combination [4.1.3, 4.2, 4.3, 4.4, 9.1, 9.2].

β_T (or β_A, β_B, β_{AB}, etc.): Column vector of regression parameters based on error plus treatment effect (or one of several factorial effects) in analysis of balanced multiple covariance [4.3].

γ (gamma): Arbitrary multiple of β [4.4].

δ (delta): Column vector of differences between means of two populations for related variables [4.5].

μ (mu): Column vector of population means for p related variables [4.5, 4.6].

σ (sigma): Column vector of components of variance [4.4.5].

τ (tau): Column vector of fixed effects of treatments [8.4].

Solutions to Odd-Numbered Exercises

CHAPTER 1

1.1.   (a) 74.7, 8.3.   (b) 0, $\sum_{i=1}^{n}(y_i-\bar{y}) = \sum_{i=1}^{n}y_i - \sum_{i=1}^{n}\bar{y} = n\bar{y}-n\bar{y} = 0$.   (c)

   $\sum_{i=1}^{9}cy_i/n = c(\sum_{i=1}^{9}y_i/n) = c\bar{y} = 10(8.3) = 83$.   (d) 719.25, 5580.09.   (e)
   99.24 in either case.

1.3.   (a) 6.765, 7, 7.5, 11, 7.816, 2.80, 0.414.   (b) 190, 176, 285.5, 523,
   13473, 116, 0.61, the median.

1.5.   (a) 0.33.   (b) 0.0275.

1.7.   (a) 0.86.   (b) 1.09, 0.6.

1.9.   (a) 50, 16.   (b) 15, 0.16.

1.11.   0.17.

1.13.   0.00856.

1.15.   0.0034.

1.17.   (a) 0.5^-.   (b) 0.683.   (c) 0.00621.

1.19.   (a) 0.5, 0.0111, 0.   (b) Increase 2.94 g.

1.21.   (a) 0.46, 1.24, 3.0, 6.5, 12.5, 21.7, 33.7, 46.9, 58.5, 65.3, 65.3,
   58.5, 46.9, 33.7, 21.7, 12.5, 6.5, 3.0, 1.24, 0.46.   (b) 0.43.

1.23. $2\bar{Y}-1$.

1.25.   (a) 10.40±0.39.   (b) 305.

1.27.   16.6±24.9.

1.29.   (a) $H{:}\mu = 40$, $\bar{H}{:}\mu < 40$; $z = -1.876$; reject H; 95% confidence that the
   true absorption rate is lower than 40 in deficient individuals. Howev-
   er, there was no control group of nondeficient individuals to ensure
   similar conditions implied for the "known" population. Reduce bias and
   experimental error by measuring absorption in the same animals before
   and after induced deficiency.   (b) $n = 99$ for $\alpha = 0.05$, $\beta = 0.2$, ignoring the control problem.

1.31.   (a) $H{:}\mu = 40$, $\bar{H}{:}\mu < 40$; $t = -2.762$; $-t_{0.01,9} = -2.821$; may not have 99%
   confidence that $\mu < 40$; no control--improve by measuring the same orga-
   nisms at both 20°C and 13°.   (b) $n = 33$, ignoring the control problem.

1.33.   (a) $H{:}\mu = 275$, $\bar{H}{:}\mu > 275$; $t = 3.133$, $t_{0.01,19} = 2.539$; yes.   (b) $n = 43$,
   55, 65 at 99%; $n = 26$, 36, 44 at 95%; $n = 19$, 28, 35 at 90%.   (c) Va-
   lidity depends on whether one merely wishes to be convinced that $\mu >$
   275 for the experimental ration or to be convinced that the "innova-
   tion" in the diet is responsible for an increase. In the latter case,

control information is required.

1.35. (a) Yes (very little evidence to reject $H: \sigma_1^2 = \sigma_2^2$). (b) 50.

1.37. (a) Yes; $z = -2.219$; $z_{0.025} = -1.96$; 95% confident $\mu_1 \neq \mu_2$. (b) 696 per group.

1.39. (a) Yes; $t = -3.024$, $-t_{0.025,8} = -2.306$. (b) 0.62.

1.41. (a) Yes; $t' = 1.856$, $t_{0.05,11.7} \approx 1.786$. (b) 35 per group if variances were equal; more (perhaps 70 per group) because variances are not equal.

1.43. (a) Yes; $q = 7.9$, $q_{0.05,5} \approx 4.8$ (Lohrding's table). (b) 25 per group if $\sigma \approx s_2$.

1.45. (a) 98.5%, approximately; $t = 2.589$. (b) 97%, approximately, for a 2-sided alternative. (c) 12 pairs. (d) $n > 100$, i.e., more than 50 per group.

1.47. (a) $z = 1.83$, $z_{0.95} = 1.654$, $z_{0.975} = 1.96$; 90% confident (but not 95%) that the procedure produces more than 50% pathogen-free pigs. (b) $q = 3.35$, $\chi_{0.10,1}^2 = 2.706$, $\chi_{0.05,1}^2 = 3.841$; same conclusion as in (a).

1.49. Yes; $q = 2.066$, $\chi_{0.3,3}^2 = 3.665$ (i.e., less than 70% confidence Poisson does *not* fit).

1.51. (a) Yes; $q = 5.741$, $\chi_{0.05,1}^2 = 3.841$. (b) 50 (25 in each group).

1.53. Hypothesis of independence of disease, inoculation, and location (infantry, supply) is rejected ($q = 211$, $\chi_{0.001,1}^2 = 10.83$); i.e., effect of inoculation on resistance to disease depends on conditions (locations) of the men. Ignoring location, one would conclude that inoculation is effective ($q = 11.83 > \chi_{0.001,1}^2 = 10.83$), but that analysis ignores the fact that most of the inoculated men enjoyed more favorable conditions. For infantrymen alone, $q = 0.05$, and for noninfantry alone, $q = 0.07$; both values are far short of significance. Very little evidence exists that inoculation is effective in either group.

1.55. (a) Sales and number of cows significantly correlated ($r_{sn} = -0.52$, $t = -2.583$), but sales and amount (kg) not significantly correlated ($r_{sk} = +0.13$, $t = 0.556$); r_{sn} negative because both price and production per cow increased (sales) while number of cows declined; r_{sk} not significant because increased production per cow and declining numbers resulted in stable quantities sold, whereas rising prices increased sales. This illustrates the fact that variables ignored (price, production per cow) can explain apparent paradoxes. (b) -0.76 to -0.10.

1.57. (a) $\hat{Y} = 100.34 + 0.716x$; 43%. (b) $H: \beta_1 = 0$; $t = 4.56$; $t_{0.005,28} = 2.763$; reject H. (c) 146.88±4.45. (d) $f = 0.36$; lack of fit not significant.

1.59. (a) Males: $\hat{Y}_2 = -1.61 + 0.0886y_1$; females; $\hat{Y}_2 = +2.13 + 0.0533y_1$. (b) Slope for males: 0.089±0.043; origin for males: -1.61±4.39; slope for females: 0.053±0.031; origin for females: +2.13±3.04. One may have 95% confidence in each case that the true parameter value lies in the interval. (c) Yes (qualified); $f = 2.722$, $f_{0.10,2,14} = 2.73$, $f_{0.05,2,14}$

= 3.74. There is a suggestion of heterogeneous regression that might be confirmed with larger samples.

CHAPTER 2

2.1. (a) f_{max} = 3.45, $f_{max,0.25,4,6}$ = 5.05; accept homogeneous variance; 3.80±0.148, 3.43±0.138, 3.60±0.160, 3.94±0.138. (b) Yes, f = 10.74, $f_{0.001,3.25}$ = 7.45. (c) $\hat{\sigma}^2$ = 0.036, with standard error 0.01; estimate of σ^2 not highly reliable. (d) 5 rats per diet.

2.3. $87.0 \leq \sigma^2 \leq 403.8$; $24.9 \leq \sigma_T^2 \leq 650.3$.

2.5. (a) Largest residual is the first count for the sixth mouse, +8.7; $e_L/\sqrt{ms_E}$ = 1.936, critical value is 2.563 for 10% significance; accept the residual as normal. (b) w = 0.988, $w_{0.50,30}$ = 0.967; normality is not rejected even at 50% significance level. (c) $11.8 \leq \sigma^2 \leq 42.1$ (within); $197 \leq \sigma_T^2 \leq 1429$ (among). (d) 15 mice per dose. (e) Estimated variation among mice (424) is more than 20 times as large as the estimated variation among counts within mice (20.2). Therefore, replication of mice per dose is much more important than replication of counts per mouse.

2.7. (a) 36% among rabbits; 64% within rabbits. (b) $0.373 \leq \sigma^2 \leq 1.044$ (within); $0.057 \leq \sigma_T^2 \leq 2.48$ (among). Components are unreliably estimated, especially the component among rabbits.

2.9. (a) Value of fish substitutes: 1 vs. 2, 3, 4, 5; coefficients {+4, −1, −1, −1, −1}. Necessity of skim milk: 2, 3 vs. 4, 5; coefficients {0, +1, +1, −1, −1}. Value of whey substitute: 2 vs. 3; coefficients {0, +1, −1, 0, 0}. Value of Promosoy without skim: 4 vs. 5; coefficients {0, 0, 0, +1, −1}. (b) 9.96±4.2, 5.38±3.75, −0.93±5.32, −5.54±5.32. (c) 18 heifers per group.

2.11. (a) Tukey's 95% significant difference is 33.89: A B̲ ̲C̲ D̲ E̲ ̲F̲. (b) Scheffé's 95% intervals are 229.3±68.8, 117.7±39.7, $\overline{67.7±39.7}$, and 17.9±39.7; D, E, and F collectively are significantly higher than the other groups; F is significantly higher than A and D but not E.

2.13. (a) Control vs. others {+5, −1, −1, −1, −1, −1}; ureas vs. ammonias {0, +3, +3, −2, −2, −2}; urea vs. urea + minerals {0, +1, −1, 0, 0, 0}; ammonia vs. ammonias⁺ {0, 0, 0, +2, −1, −1}; effect of molasses {0, 0, 0, 0, +1, −1}. (b) $\hat{\sigma} \approx (36-12)/6 = 4$; Δ = 2.81. (c) Bonferroni, 3.72; Scheffé, 4.82; Tukey, 4.13; Dunnett, 3.62.

2.15. (a) Determinations: $\hat{\sigma}_D^2$ = 0.0080 (6.2%); samples: $\hat{\sigma}_S^2$ = 0.0318 (24.8%); bulls: $\hat{\sigma}_B^2$ = 0.0883 (68.9%). (b) $0.0058 \leq \sigma_D^2 \leq 0.0118$; $0.0205 \leq \sigma_S^2 \leq 0.0561$; $0.0479 \leq \sigma_B^2 \leq 0.2157$. (c) Repeatability of assay technique = 0.94; repeatability of blood samples = 0.70. (d) 1.7 or 2 determinations per sample. (e) 7 samples per bull. (f) 233%.

2.17. (a) 1.4%; 14.5%; 84.1%. (b) No; Satterthwaite's procedure: f = 1.1,

$f_{0.25,2,4.55} \approx 1.92$. (c) $57 \le \sigma^2 \le 9619$; $932 \le \sigma_U^2 \le 2046$. (d) 3.93 or 4 samples per subject. (e) 9 subjects per treatment group.

2.19. (a) Bartlett's test: $q = 1.255$, $\chi^2_{0.3,2} = 2.408$; accept homogeneous variance among samples. Treatments (Satterthwaite's procedure): $f = 1.60$, $f_{0.10,2,15.7} \approx 2.68$; not even 90% confident that treatments differ. (b) 0.0148; 0.0047. (c) 2.28±0.161; 2.39±0.193; 2.52±0.131. (d) 68.4%. (e) 11 samples per treatment.

2.21. (a) $Y_{ijk} = \mu + \alpha_i + B_{(i)j} + \gamma_k + (\alpha\gamma)_{ik} + (B\gamma)_{(i)jk} + E_{(ijk)}$, where Y_{ijk} = response for one pig; μ = overall mean; α_i = fixed effect of ith breed; $B_{(i)j}$ = random effect of jth litter from ith breed; γ_k = fixed effect of kth diet; $(\alpha\gamma)_{ik}$ and $(B\gamma)_{(i)jk}$ are interactions of diets with breeds and litters; $E_{(ijk)}$ = experimental error (unnamed effects peculiar to one pig), not separable from $(B\gamma)_{(i)jk}$ in this experiment. (b) ss_{BC} =

$$(\sum_{i=1}^{3} \sum_{j=1}^{15} \sum_{k=1}^{6} y_{ijk}^2) - (\sum_{i=1}^{3} \sum_{j=1}^{15} y_{ij.}^2/6) - (\sum_{i=1}^{3} \sum_{k=1}^{6} y_{i.k}^2/15) + (\sum_{i=1}^{3} y_{i..}^2/90).$$

(c)

Source of Variation	df	$E[MS]$
Breeds (A)	2	$\sigma^2 + 6\sigma_B^2 + 90\Sigma\alpha_i^2/2$
Litters/breeds (B/A)	42	$\sigma^2 + 6\sigma_B^2$
Diets (C)	5	$\sigma^2 + \sigma_{BC}^2 + 45\Sigma\gamma_k^2/5$
AC	10	$\sigma^2 + \sigma_{BC}^2 + 18\Sigma\Sigma(\alpha\gamma)_{ik}^2/10$
$[BC+E]$	210	$\sigma^2 + \sigma_{BC}^2$

(d) $H:\alpha_i = 0$: $f = ms_A/ms_B$ vs. $f_{\alpha,2,42}$; $H:\gamma_k = 0$: $f = ms_C/ms_{BC}$ vs. $f_{\alpha,5,210}$; $H:(\alpha\gamma)_{ik} = 0$: $f = ms_{AC}/ms_{BC}$ vs. $f_{\alpha,10,210}$.

2.23. (a) $Y_{ijkl} = \mu + \alpha_i + B_{(i)j} + \gamma_k + \delta_l + (\alpha\gamma)_{ik} + (\alpha\delta)_{il} + (B\gamma)_{(i)jk} + (B\delta)_{(i)jl} + (\gamma\delta)_{kl} + (\alpha\gamma\delta)_{ikl} + (B\gamma\delta)_{(i)jkl} + E_{(ijkl)}$; where Y_{ijkl} = response for one rat; μ = overall mean; α_i = fixed effect of ith treatment; $B_{(i)j}$ = random effect of jth litter assigned to ith treatment; γ_k = fixed effect of kth sex; δ_l = fixed effect of lth diet; $(\alpha\gamma)_{ik}$, $(\alpha\delta)_{il}$, $(\gamma\delta)_{kl}$, and $(\alpha\gamma\delta)_{ikl}$ are interactions among treatments, sexes, and diets; $(B\gamma)_{(i)jk}$, $(B\delta)_{(i)jl}$, and $(B\gamma\delta)_{(i)jkl}$ are interactions of litters with sexes and diets (like-ly negligible); and $E_{(ijkl)}$ = experimental error (unnamed effects pe-culiar to one rat), not separable from $(B\gamma\delta)_{(i)jkl}$ in this experiment.

(b) $ss_{AD} = (\sum_{i=1}^{2} \sum_{l=1}^{3} y_{i..l}^{2}/12) - (\sum_{i=1}^{2} y_{i...}^{2}/36) - (\sum_{l=1}^{3} y_{...l}^{2}/24) + (y_{....}^{2}/72)$

$ss_{BD} = (\sum_{i=1}^{2} \sum_{j=1}^{6} \sum_{l=1}^{3} y_{ij.l}^{2}/2) - (\sum_{i=1}^{2} \sum_{j=1}^{6} y_{ij..}^{2}/6) - (\sum_{i=1}^{2} \sum_{l=1}^{3} y_{i..l}^{2}/12)$

$\qquad + (\sum_{i=1}^{2} y_{i...}^{2}/36)$

$ss_{CD} = (\sum_{k=1}^{2} \sum_{l=1}^{3} y_{..kl}^{2}/12) - (\sum_{k=1}^{2} y_{..k.}^{2}/36) - (\sum_{l=1}^{3} y_{...l}^{2}/24) + (y_{....}^{2}/72)$

(c)

Source of Variation	df	$E[MS]$
Treatments (A)	1	$\sigma^2 + 6\sigma_B^2 + 36\Sigma\alpha_i^2$
Litters/treatments (B/A)	10	$\sigma^2 + 6\sigma_B^2$
Sexes (C)	1	$\sigma^2 + 3\sigma_{BC}^2 + 36\Sigma\gamma_k^2$
Diets (D)	2	$\sigma^2 + 2\sigma_{BD}^2 + 24\Sigma\delta_l^2/2$
AC	1	$\sigma^2 + 3\sigma_{BC}^2 + 18\Sigma\Sigma(\alpha\gamma)_{ik}^2$
AD	2	$\sigma^2 + 2\sigma_{BD}^2 + 12\Sigma\Sigma(\alpha\delta)_{il}^2/2$
BC	10	$\sigma^2 + 3\sigma_{BC}^2$
BD	20	$\sigma^2 + 2\sigma_{BD}^2$
CD	2	$\sigma^2 + \sigma_{BCD}^2 + 12\Sigma\Sigma(\gamma\delta)_{kl}^2/2$
ACD	2	$\sigma^2 + \sigma_{BCD}^2 + 6\Sigma\Sigma\Sigma(\alpha\gamma\delta)_{ikl}^2/2$
$[BCD+E]$	20	$\sigma^2 + \sigma_{BCD}^2$

(d) $H:\alpha_i = 0$: $f = ms_A/ms_B$ vs. $f_{\alpha,1,10}$; $H:\gamma_k = 0$: $f = ms_C/ms_{BC}$ vs. $f_{\alpha,1,10}$; $H:\delta = 0$: $f = ms_D/ms_{BD}$ vs. $f_{\alpha,2,20}$. (e) To separate components, one must assume that σ^2_{BCD} is negligible (interaction of litters, sexes, and diets). Then, $\hat{\sigma}^2 = ms_{BCD}$ (error); $\hat{\sigma}^2_{BD} = (ms_{BD} - ms_{BCD})/2$; $\hat{\sigma}^2_{BC} = (ms_{BC} - ms_{BCD})/3$; $\hat{\sigma}^2_B = (ms_B - ms_{BCD})/6$.

2.25.   (a)

Source of Variation	df	ss	ms	$E[MS]$
Vaccines (A)	3	39.82	13.27	$\sigma^2 + 18\Sigma\alpha_i^2/3$
Additive (B)	5	33.07	6.61	$\sigma^2 + 12\Sigma\beta_j^2/5$
AB	15	37.76	2.52	$\sigma^2 + 3\Sigma\Sigma(\alpha\beta)_{ij}^2/15$
Error	48	151.33	3.15	σ^2

(b) $f = 0.8$, $f_{0.50,15,48} = 0.97$; not even 50% confidence that interaction exists. Differences among vaccines are likely to be consistent from level to level of additive. (c) A: $f = 4.21$, $f_{0.05,3,48} = 2.8$; vaccines differ in effectiveness (95% confidence); B: $f = 2.10$, $f_{0.10,5,48} = 1.98$, $f_{0.05,5,48} = 2.41$; one may have 90%, but not 95%, confidence that effectiveness of vaccines is influenced by amount of additive. If vaccines had random effects, then $E[MS_B] = \sigma^2 + 3\sigma^2_{AB} + 12\Sigma\beta_j^2/5$ and $E[MS_{AB}] = \sigma^2 + 3\sigma^2_{AB}$, so $f = 2.62$, $f_{0.10,5,15} = 2.27$, $f_{0.05,5,15} = 2.90$. In this case inference is affected very little. (d) Standard errors for vaccine, additive, and combination means: ±0.42, ±0.51, ±1.02. (e) Tukey's 95% significant difference = 1.58; 4   2   1   3; vaccine 3 more effective than 4, vaccine 1 nearly so at 95%. (f) 7 serum pools per combination. (g) Linear, $f = 10.04$; quadratic, $f = 0.17$; cubic, $f = 0.20$; $f_{0.01,1,48} = 7.22$. Linear parameter clearly important; little evidence of need for nonlinear parameters in the range of additive studied. (h) $\hat{Y} = 5.25 + 0.281\xi'_{j1}$ or $Y = 4.76 + 0.14x = 5.53$ for $x = 5.5$.

2.27. (a) Interaction: $f = 0.87$ (not significant); treatments: $f = 4.34$, $f_{0.05,3,24} = 3.01$; dose: $f = 34.00$, $f_{0.001,1,24} = 14.0$; treatments differ. Dose effect is highly significant; little evidence for interaction. (b) Spacing (4/7, 5/7, 7/7) corresponds to (0, 1, 3); control vs. others: $t = 2.46$; linear: $t = -2.60$; quadratic: $t = -0.43$, $t_{0.025,24} = 2.064$. Effect of restricted feeding in elevating levels of FSH-releasing factor appear to be linear in the range studied. (c) A: ±5.9; B: ±4.2. (d) 11.5±17.2; 22.0±17.2. (e) 10 per cell.

2.29. (a) Expected mean squares are divided by the number of observations per cell relative to what they would be in the full analysis. Also, ABC interaction is not separable from experimental error. (b) $(\hat{\sigma}^2/15) + \hat{\sigma}^2_{ABC}$

$= 321$. If $\sigma^2_{ABC} = 0$, then $\hat\sigma^2 = 4815$ (experimental error). $\hat\sigma^2_A = 8.9$,
$\hat\sigma^2_B = 209.0$, $\hat\sigma^2_C = -61.1$ (use zero as best estimate), $\hat\sigma^2_{AB} = 9.0$, $\hat\sigma^2_{AC}$
$= 135.8$, $\hat\sigma^2_{BC} = 176.9$; 95% confidence intervals: $3413 \le \sigma^2 \le 7307$,

$0.1 \le \sigma^2_A \le 641$, $46 \le \sigma^2_B \le 6663$, not possible for σ^2_C, $0.2 \le \sigma^2_{AB} \le 1026$,

$58 \le \sigma^2_{AC} \le 547$, $68 \le \sigma^2_{BC} \le 1567$. (c) 0.009; 0.22. (d) $f = 3.96$,
$f_{0.10,3,6} = 3.29$, $f_{0.05,3,6} = 4.76$; 90% (but not 95%) confidence that
variation among locations is not zero. (e) Validity of tests involving
years (factor C) depends on the truth of assuming $\sigma^2_{ABC} = 0$. AB interaction: $f = 1.08$, $f_{0.05,27,54} = 1.70$ (no strain-location interaction).
AC interaction: $f = 2.69$, $f_{0.05,18,54} = 1.82$ (strains not consistent
from year to year). BC interaction: $f = 6.5$, $f_{0.001,6,54} = 4.48$ (location effects differ from year to year). C (years): $f = 0.59$ (year
effects not significant on the average). Because of AC and BC interaction, compare strains within years: Tukey's 95% significant difference
$= 41.9$; no significant differences in year 1. Strain 3 significantly
lower than strains 1, 2, 4, 6, and 8 in year 2. Strain 8 significantly
lower than strains 4 and 9 in year 3 (over all years, 3 is lower than
1, 4, and 9). Absence of AB interaction suggests that different
strains need not be developed for different regions, but comparisons of
strains within any one year are not reliable indicators of long-range
average value of the strain.

2.31. (a)

Source of Variation	df	ss	ms	f ratio
Age (A)	1	1.477	1.477	26.38
Days (B)	5	0.607	0.121	2.16
AB	5	0.517	0.103	1.84
Error	24	1.349	0.056	

$f_{0.001,1,24} = 14.0$, $f_{0.10,5,24} = 2.10$, $f_{0.05,5,24} = 2.62$. Effects of
age are highly significant; effects of days of ad libitum feeding may
be significant (90% confidence), but there is not strong evidence for
interaction. (b) Standard errors of cell means:

Age	Days=45	90	135	180	225	270
2	1.14±0.12	1.08±0.14	1.14±0.11	1.10±0.12	1.13±0.14	1.17±0.12
>2	1.56±0.17	1.55±0.14	1.19±0.24	1.65±0.17	1.38±0.14	2.10±0.17

Plot of cell means against days indicates that persistency is not related to length of ad libitum feeding for 2-year-old cows. For older cows feeding 270 days may be the best policy, although results are not clear because of large standard errors. Scheffé's 95% interval for {(45, 90, 135, 180, 225) vs. 270 days} is -3.17±3.33, i.e., evidence is almost sufficient (but not quite) for 95% confidence that mean persistency is lower for ad libitum feeding 45 to 225 days than for 270 days.

2.33. (a) $Y_{ijk} = \mu + \alpha_i + B_{(i)j} + \gamma_k + (\alpha\gamma)_{ik} + (B\gamma)_{(i)jk} + E_{(ijk)}$, where Y = response at one quarter of one silo; μ = overall mean; α = treatment effect; B = random effect of a silo (within treatment); γ = quarter effect; $\alpha\gamma$ and $B\gamma$ are interactions of treatment and silo with quarter; E = experimental error, inseparable from $B\gamma$ in this experiment.

Source of Variation	df	ms	E[MS]
Treatments (A)	1	455.91	$\sigma^2 + 4\sigma_B^2 + 44\Sigma\alpha_i^2$
Silos/treatments (B/A)	20	2.47	$\sigma^2 + 4\sigma_B^2$
Quarters (C)	3	1.04	$\sigma^2 + \sigma_{BC}^2 + 22\Sigma\gamma_k^2/3$
AC	3	0.13	$\sigma^2 + \sigma_{BC}^2 + 11\Sigma\Sigma(\alpha\gamma)_{ik}^2/3$
$[BC+E]$	60	1.32	$\sigma^2 + \sigma_{BC}^2$

$H:\alpha_i = 0$, $f = 184.6$, $f_{0.001,1,20} = 14.8$; highly confident urea treatment increases crude protein, 99% CI $(\alpha_1 - \alpha_2) = 4.55 \pm 0.95\%$; $H:\gamma_k = 0$, $f = 0.788$, little evidence of real differences among quarters on the average; $H:(\alpha\gamma)_{ik} = 0$, $f = 0.098$, little evidence of interaction between treatments and quarters. (b) Nonrepeatable effects of quarters should be nested within silos. Quarter effects could still be tested if 2 or more samples were taken from each quarter, i.e., if a 4-stage nested model is used: $Y_{ijkl} = \mu + \alpha_i + B_{(i)j} + \gamma_{(ij)k} + E_{(ijk)l}$ ($l = 1,2$). (Note: Substitute C for γ if quarter effects are considered random.)

Source of Variation	df	E[MS]
Treatments (A)	1	$\sigma^2 + 8\sigma_B^2 + 88\Sigma\alpha_i^2$
Silos/treatments (B/A)	20	$\sigma^2 + 8\sigma_B^2$
Quarters/silos (C/B)	66	$\sigma^2 + 2\Sigma\Sigma\Sigma\gamma_{(ij)k}^2/66$
Error	88	σ^2

(Note: If quarter effects are considered random, $2\sigma_C^2$ will appear in all expected mean squares except error.)

$$ss_A = \left(\sum_{i=1}^{2} y_{i\ldots}^2 / 88\right) - \left(y_{\ldots\ldots}^2 / 176\right) \qquad ss_B = \left(\sum_{i=1}^{2} \sum_{j=1}^{11} y_{ij\ldots}^2 / 8\right)$$

$$-\left(\sum_{i=1}^{2} y_{i\ldots}^2 / 88\right) \qquad ss_C = \left(\sum_{i=1}^{2} \sum_{j=1}^{11} \sum_{k=1}^{4} y_{ijk\ldots}^2 / 2\right) - \left(\sum_{i=1}^{2} \sum_{j=1}^{11} y_{ij\ldots}^2 / 8\right)$$

$$ss_E = ss_y - ss_A - ss_B - ss_C$$

where $ss_y = \sum_{i=1}^{2} \sum_{j=1}^{11} \sum_{k=1}^{4} \sum_{l=1}^{2} y_{ijk\ell}^2 - \left(y_{\ldots\ldots}^2 / 176\right)$.

CHAPTER 3

3.1. (a) Yes, $f = 19.42$. (b) No, $b = 1.104$ for 0.025 mL; $b = 0.957$ for 0.10 mL; $f = 1.156$.

3.3. (a) No, slopes are somewhat more alike than one would expect by chance ($1/f > f_{0.25,36,1}$). (b) 0.915 ± 0.078. (c) Slope significantly different from zero ($f = 128.9$); efficiency: 233%. (d) Yes, $f = 35.37$.

3.5. (a) $f_{max} = 1.47$, $f_{max,0.25,4,14} \simeq 2.79$; accept homogeneous variance.
(b) $f = 2.29$, $f_{0.10,6,55} = 1.89$; one or more of the assumptions may be suspect; equal slopes: $f = 0.373$ (not significant); linear regression of means: $f = 5.69$, $f_{0.01,2,58} = 5.00$; not linear; coordinate for means of the first sire group is out of line; validity of covariance adjustment is unclear. (c) Unadjusted means: 0.660, 0.606, 0.584, 0.590; only 1 vs. 3 is significant using Tukey's test at 90%; adjusted means: 0.649, 0.629, 0.567, 0.596; rankings unchanged by adjustment, but 1 vs. 3 is significant at 95% using Tukey's test as an approximation with effective error. (d) 0.649 ± 0.036; 0.629 ± 0.037; 0.567 ± 0.035; 0.596 ± 0.038. (e) 129% (ignoring the problem of nonlinear means).

3.7. (a) Yes; Bartlett's test of heterogeneous variance is not significant ($q = 1.395 < \chi_{0.3,3}^2 = 3.665$). (b) No ($f = 0.835 < f_{0.25,3,80} = 1.40$); 0.623 ± 0.165 for common slope. (c) Sex effect is significant at 95% level ($f = 5.73$); male age ($f = 0.622$) and female age ($f = 0.138$) are not significant even at 75% confidence.

CHAPTER 4

4.1. (a) $\begin{bmatrix} 17 & 0 & 7 \\ -16 & 12 & -8 \\ 19 & 7 & 7 \end{bmatrix}$ (b) $\begin{bmatrix} 14 & -1 \\ 22 & 22 \end{bmatrix}$ (c) $\begin{bmatrix} 5 & 5 \\ 1 & -7 \\ 6 & 5 \end{bmatrix}$ (d) $\begin{bmatrix} 1 & -1 \\ -1 & -1 \\ 4 & -3 \end{bmatrix}$

(e) $\begin{bmatrix} -7 & -13 \\ -5 & 11 \\ 0 & -19 \end{bmatrix}$ (f) $\begin{bmatrix} 9 & 7 \\ 7 & 29 \end{bmatrix}$ (g) $\begin{bmatrix} 18 & -5 & 8 \\ -9 & 11 & -5 \\ 19 & -10 & 9 \end{bmatrix}$ (h) $\begin{bmatrix} 23 & 6 \\ 29 & 51 \end{bmatrix}$ (i) $\begin{bmatrix} 41 & 79 \\ -28 & -116 \\ 52 & 64 \end{bmatrix}$

4.3. $x = 13/77$, $y = 65/77$, $z = 21/77$.

4.5. (a) $\hat{\beta}_0 = 15.36 \pm 0.36$, $\hat{\beta}_1 = -0.627 \pm 0.073$, $\hat{\beta}_2 = 0.0577 \pm 0.0272$. (b) t_1 $= -8.657$, $t_2 = 2.121$, $\pm t_{0.0025,21} = 2.080$; $R^2 = 0.816$. (c) $\hat{Y} = 11.5$ ± 1.57.

4.7. $\hat{Y} = 3.13 \pm 0.645$ (95%).

4.9. (a) $\hat{Y} = -8.8 + 24.7x - 13.6x^2$. (b) $R^2 = 0.90$. (c) $\hat{Y} = 1.79 \pm 0.06$ for $x =$ 0.7. (d) $\hat{x}_{opt} = 0.91$.

4.11. (a) Adjusted means (by dose): 38.64(0.10); 49.72(0.125); 74.80(0.20); 86.40(0.25). (b) $f = 2430.3/14.52 = 167.4 \gg f_{0.001,3,14} = 9.73$; reject $H:\tau_i = 0$.

4.13. Sires: $\hat{\sigma}_A^2 = 343,839$; herds: $\hat{\sigma}_B^2 = 2,134,026$; herd-season interaction: $\hat{\sigma}_{BC}^2 = 324,606$; error: $\hat{\sigma}_E^2 = 1,796,716$.

4.15. (a) $q = 24.6 > \chi_{0.01,6}^2 = 16.8$. Covariance matrices are not equal. (b) Differences (control minus treated) of random pairs (1,3), (2,4), (3,6), (4,9), (5,8), (6,1), (7,10), (8,11), (9,7), (10,12), treated numbers 2 and 5 discarded; 95% CI: -0.173 ± 0.216, 0.084 ± 0.151, 0.004 ± 0.034.

4.17. $U = 0.608 < U_{0.05,3,3,41} = 0.656$ (groups differ significantly).

CHAPTER 5

5.1. (a) $f = 48.52$, $f_{0.001,2,10} = 14.9$; treatments differ. (b) Linear: f $= 94.92$, $f_{0.001,1,10} = 21.0$ (significant); quadratic: $f = 2.127$, $f_{0.10,1,10} = 3.29$ (not significant). (c) 251%.

5.3. (a) Sum of squares for nonadditivity $= 1.4 \times 10^{-5}$, $f \ll 1$ (nonsignificant); dependence of variance on level of response: $f = 1.334 <$ $f_{0.25,1,12} = 1.46$ (not significant). (b) Overall treatment differences: $f = 17.8$, $f_{0.001,2,12} = 13.0$ (highly significant). Dunnett: drug 1, $t_d = 4.623$; drug 2, $t_d = -0.98$; $t_{d,0.01,2,12} = 3.39$; drug 1 differs from control; drug 2 not significant. (c) 40 blocks (litters). (d) 1070%. (e) $\hat{y}_{24} = 5.4$, bias in $ss_T = 0.33$.

5.5. (a) 100%; blocking no more efficient than completely randomized design. (b) None, because σ_{TD}^2 appears in $E[MS_T]$ as well as in $E[MS_E]$. (c) 124%; an unsuspected interaction can mask the efficiency of blocking. (d) 25 litters. (e) Correlation within treatment groups $= 0.44$.

5.7. (a) $\hat{y}_{126} = 0.133$; bias $= 10^{-5}$ (ignore bias). (b) Organisms: $f = 0.03$;

sterilization: $f = 11.18$; interaction: $f = 7.12$, $f_{0.05,1,26} = 4.23$, $f_{0.01,1,26} = 7.72$; interaction is significant and average sterilization effect is highly significant; little evidence of difference between organisms. Sterilization compared separately for each organism is significant for *N. sitophilus* ($t = 4.302$) but not for *S. carlsbergensis* ($t = 0.483$).

CHAPTER 6

6.1. (a) 79.7, 94.5, 77.3, 82.7, 73.4, 87.7, 88.0, 96.3, 94.9, 95.0, 64.3, 84.0, 67.2, 97.8, 79.8, 90.9, 79.2, 93.6, 69.3, 83.1, 90.8. (b) Tukey's 95% significant difference = 15.55.
   14, 8 > 15, 1, 17, 3, 5, 19, 13, 11; 10, 9 > 17, 3, 5, 19, 13, 11; 2, 18 > 3, 5, 19, 13, 11; 16, 21 > 5, 19, 13, 11; 7, 6 > 19, 13, 11; 12, 20 > 13, 11; 4 > 11
   (c) 79.2, 94.6, 76.9, 82.6, 73.8, 87.9, 87.6, 96.2, 95.2, 95.6, 63.8, 83.8, 67.6, 98.5, 79.6, 90.9, 78.6, 94.0, 69.5, 82.6, 90.8. None of the combined estimates differ from corresponding intrablock estimates by more than 0.7 (< 1% of the means). (d) 101.7%.

6.3. (a) 30. (b) $k = 4$. (c) $k = 5$; only 6 litters required, minimum efficiency = 0.96 vs. 0.6, 0.8 for $k = 2, 3$. (d) 0.042 for $k = 5$.

6.5. Try different defining contrasts until one is found that produces zero residue for all treatment combinations listed. For example, $ABCD$ is not the defining contrast because d, or {0001}, produces $R = 1$ for $\sum_{i=1}^{4} x_i / 2$. The defining contrast is ABC because all combinations listed produce $R = 0$ for $\sum_{i=1}^{3} x_i / 2$. Sources of variation (df) are: replicates (3), blocks/replicates (ABC) (4), main effects (4), 2-factor interactions (6), 3-factor interactions (3), $ABCD$ (1), error (42).

6.7. (a) $\hat{\Delta}_A = 0.62$, $\hat{\Delta}_B = 1.88$, $\hat{\Delta}_{AB} = -9.38$, $\hat{\Delta}_C = 6.88$, $\hat{\Delta}_{AC} = 0.62$, $\hat{\Delta}_{BC} = -0.62$, $\hat{\Delta}_{ABC} = -1.88$ (confounded with hospitals); significant effects must exceed 2.80 (95%) or 4.24 (99%); psychotherapy (C) is significant as is interaction of the 2 drugs (AB); apparently, the effect of psychotherapy is independent of the effects of the 2 drugs. (b) Effect of A when B is present, absent: -8.76, $+10.00$; effect of B when A is present, absent: -7.50, $+11.26$; 95% significant effect must exceed 5.99; both drugs have effect but should not be given together. Difference between A alone and B alone is -1.26 (not significant).

6.9. (a) Defining contrasts: $ABCD$, $CDEF$, ACF; automatically confounded: $ABEF$, BDF, ADE, BCE.

Block 1	2	3	4	5	6	7	8
(1)	ab	e	f	b	a	d	c
ace	bce	ac	acef	abce	ce	acde	ae
adf	bdf	adef	ad	abdf	df	af	acdf
bcf	acf	bcef	bc	cf	abcf	bcdf	bf
bde	ade	bd	bdef	de	abde	be	bcde
abcd	cd	abcde	abcdf	acd	bcd	abc	abd
abef	ef	abf	abe	aef	bef	abdef	abcef
cdef	abcdef	cdf	cde	bcdef	acdef	cef	def

(b) Sources of variation (df): blocks (7), main effects (6), 2-factor interactions (15), error = unconfounded higher order interactions (35).

6.11. (a) Fundamental identity: $I = ABCE = ABDF = CDEF$; defining contrast: ACF (blocking); Yates's analysis with e and f ignored: 95% significant effect must exceed 50.04; significant effects and aliases are $A = BCE = BDF = ACDEF$, $AB = CE = DF = ABCDEF$, $BC = AE = ACDF = BDEF$, and $ABC = E = CDF = ABDEF$. Apparently there is significance of A and E, which have 3-factor aliases. Two-factor interactions are too entangled for interpretation. (b) Main effects of B, C, D, and F may be small, but interactions with A and E may be important. Try half-replicate in 2 blocks (16 dogs of each sex) for resolution VI; or to estimate 3-factor interactions, use a full replicate in 2 blocks.

6.13. $I = (ABCD^2)$; defining relation (blocking): $(AC^2) = (AB^2D) = (BC^2D^2)$.

Block 1	2	3
0000	0022	0011
0101	0120	0112
0202	0221	0210
1012	1001	1020
1110	1102	1121
1211	1200	1222
2021	2010	2002
2122	2111	2100
2220	2212	2201

Aliases: $A = (AB^2C^2D) = (BCD^2)$, $B = (AB^2CD^2) = (ACD^2)$, $C = (ABC^2D^2) = (ABD^2)$, $D = (ABC) = (ABCD)$, $(BC) = (AB^2C^2D^2) = (AD^2)$, $(BC^2) = (AB^2D^2) = (AC^2D^2)$, $(BD) = (AB^2C) = (ACD)$, $(BD^2) = (AB^2CD) = (AC)$, $(CD) = (ABC^2) = (ABD)$, $(CD^2) = (ABC^2D) = (AB)$.

Sources of variation (df): blocks (2); main effects (8); BC, BD, CD, (12); error $[(AB^2)+(AD)]$ (4).

6.15. Let $I = (ABCDEF)$. Use only combinations that produce zero residues for $\sum_{i=1}^{6} x_i/3$, i.e., combinations for vector sums 0, 3, 6, 9, or 12 before modulation. Only the combination (000000) provides sum 0, (111111) provides sum 6, and only (222222) provides sum 12. Other combinations providing sums 3, 6, 9 may be found by enumerating combinations as follows:

Vector Sum	Number of Factors at Level			Number of Trt. Combinations
	0	1	2	
3	3	3	0	$C(6,3) = 20$
	4	1	1	$C(6,4)C(2,1) = 30$
6	1	4	1	$C(6,4)C(2,1) = 30$
	2	2	2	$C(6,2)C(4,2) = 90$
	3	0	3	$C(6,3) = 20$
9	0	3	3	$C(6,3) = 20$
	1	1	4	$C(6,4)C(2,1) = 30$

6.17. Let $I = (AB^2D^2) = (BCD^2) = (ACD) = (ABC^2)$ to ensure inclusion of (1212). The 8 aliases of each main effect involve at least 2 factors each. Design: $R = 0$ for $(x_1+2x_2+2x_4)/3$, $R = 1$ for $(x_2+x_3+2x_4)/3$.

$$(1/9)3^4: \quad \begin{array}{lll} 0010 & 1021 & 2002 \\ 0122 & 1100 & 2111 \\ 0201 & 1212 & 2220 \end{array}$$

6.19. Let $I = (A'B''C'''D') = (A'B'''C'D'') = (A'B'C') = (A'''B'C''D''') = (B'''C''D') = (A'''B'D'') = (A''C'''D''')$. No main effects have components of other main effects as aliases. Two-factor aliases of main effects: $A' = (B'C'')$, $A'' = (C''D''')$, $A''' = (B''D'')$, $B' = (A'C')$, $B'' = (A'''D'')$, $B''' = (C''D')$, $C' = (A'B')$, $C'' = (B'''D'')$, $C''' = (A''D''')$, $D' = (B'''C'')$, $D'' = (A'''B'')$, and $D''' = (A''C''')$. Design pseudofactors: $R = 0$ for $(x_1+x_3+x_4+x_6+x_7)/2$, $(x_1+x_2+x_4+x_5+x_7+x_8)/2$, and $(x_1+x_3+x_5)/2$.

$$(1/8)4^4: \quad \begin{array}{llll} 0000 & 3200 & 3120 & 0320 \\ 1001 & 2201 & 2121 & 1321 \\ 1110 & 2310 & 2030 & 1230 \\ 0111 & 3311 & 3031 & 0231 \\ 0102 & 3302 & 3022 & 0222 \\ 1103 & 2303 & 2023 & 1223 \\ 1012 & 2212 & 2132 & 1332 \\ 0013 & 3213 & 3133 & 0333 \end{array}$$

6.21. Let $I = (AB^2C^3D) = (BCD) = (AB^3C^2) = (ACD^3) = (ABD^2)$. Two-factor aliases of main effects: $A = (BC^3) = (BD^2) = (CD^3)$, $B = (AC^2) = (AD^2) = (CD)$, $C = (AB^3) = (AD^3) = (BD)$, $D = (AB) = (AC) = (BC)$. Design: u_0

for $(u_{x_1} + u_2 u_{x_2} + u_3 u_{x_3} + u_{x_4})$ and $(u_{x_2} + u_{x_3} + u_{x_4})$.

$(1/16)4^4$:

0000	1033	2011	3022
0123	1110	2132	3101
0231	1202	2220	3213
0312	1321	2303	3330

6.23. Let defining relations be AB (1 df) and (CD^2E^2) (2 df), with generalized interaction $(ABCD^2E^2)$ (2 df). Six blocks according to combinations {00, 01, 02, 10, 11, 12} of residues $R_1 = 0, 1$ and $R_2 = 0, 1, 2$ from $(x_1+x_2)/2$ and $(x_3+2x_4+2x_5)/3$, respectively. In each of the first 3 blocks, $(x_1, x_2) = (00)$ for 9 treatment combinations and $(x_1, x_2) = (11)$ for the remaining 9 combinations. In the last 3 blocks, the corresponding results will be (01) and (10). Within each group of 9 treatment combinations in a half-block, x_3 and x_4 can be written in regular order, {00, 01, 02, 10, ..., 22}. Finally, determine x_5 in each case such that the proper value of R_2 is obtained. Then the order of proper values of x_5 is the same in each half of the same block. For example, in blocks 1 and 4 the order is {0, 2, 1, 1, 0, 2, 2, 1, 0}.

CHAPTER 7

7.1. (a) 65–70% confidence. (b) 56.5±16.7, 92.9±16.7, 106.4±16.7. (c) ss_{NA} = 746, $f = 0.807$ (not significant). Assumption that rows and columns do not interact with each other or with treatments appears satisfactory. Therefore, Latin squares should be valid designs. (d) Three balanced sets of 3 squares (81 heifers in all) should provide approximately 0.94 power.

7.3. (a) Yes, animal fat diet (one-sided Dunnett test: $t_D = 7.10$). (b) Efficiency of Latin squares ($\nu_1 = 16$) to completely randomized ($\nu_2 = 26$) is 341%. Yes, body weight is more important than litters with respect to efficiency, because the mean square is more than 3 times as large. (c) Yes; $m = 2$ squares minimum.

7.5. (1) Latin square (7 × 7) with rows = litters (6 df), columns = weight groups (6 df), treatments (6 df), and error (30 df); (2) RCBD with blocks = litters (6 df), covariate = weight (1 df), treatments (6 df), and error (35 df); (3) BIBD ($k = 3$ rats) with 14 blocks (13 df), treatments (6 df), error (22 df), and efficiency $E = (t\lambda/kr) = (7)(2)/(3)(6)$ = 0.78; (4) BIBD ($k = 2$ rats) with 21 blocks (20 df), treatments (6 df), error (15 df), and efficiency $E = (7)(1)/(2)(6) = 0.58$. The BIB designs would have to reduce error variation considerably to compensate for low efficiencies relative to complete blocks based on weight (columns in the Latin square) and have only 22 or 15 df for error (vs. 30 for the Latin square). However, the RCB design with covariance has 35 df for error and could be more effective than the Latin square if the relation between weight and response is strong (and linear) or if the weight groups of the Latin square are not very homogeneous because of large differences in average weight from litter to litter.

7.7.   (a)

Source of Variation	df	ss	ms
Rows	7	618.86	88.41
Columns	7	396.86	56.69
Treatments	7	19,849.86	2,835.69
Preparations	3	74.29	24.76[*]
Doses	1	19,775.39	19,775.39[*]
Interaction	3	0.18	0.06
Error	42	128.15	3.05

[*]$P < 0.01$.

(b) Efficiency relative to completely randomized is 606%.

7.9.   (a)

Stage-of-Lactation Groups	Genetic Groups 1	2	3
1	A	B	C
2	B	C	D
3	C	D	A
4	D	A	B

(b) $t = 4$ treatments, $k = 3$ (row size), $r = 3$ replications, and $\lambda = 2$ comparisons of each pair of treatments in the same row; $E = (t\lambda/kr) = 0.89$.

7.11.   (1) Double confounding: 2 repetitions of a pair of 4 × 4 Latin squares, each square containing 4 treatment combinations based on ABC as a defining contrast, with squares (3 df), rows = age/square (12 df), columns = weight/square (12 df), main effects (3 df), 2-factor interactions (3 df), and error (30 df); (2) 8 × 8 Latin square: rows = age (7 df), columns = weight (7 df), main effects (3 df), interactions (4 df), and error (42 df); (3) RCBD: 8 blocks (7 df), main effects (3 df), interactions (4 df), and error (49 df); (4) incomplete blocks: 8 replicates (7 df), each having 2 blocks of 4 pigs (blocks/replicates = ABC, 8 df), main effects (3 df), and 2-factor interactions (3 df). Block designs do not require zero interaction of age with weight as do Latin squares for valid inference about treatments. Blocking on both variables jointly could be effective, especially with incomplete blocks of 4 animals. The 8 × 8 Latin square is subject to the greater risk of treatment bias from age-weight interaction, and blocking on weight may be less effective than in block designs if there is much variation in age as the traits are highly correlated. The design with double confounding should incur smaller risk of bias and be most effective in blocking by weight because there would be 16 age classes (instead of 8) with only 4 ages per weight class. That design may be the most effective for the experiment.

CHAPTER 8

8.1.   (a) $f = 19.1 > f_{0.01,2,8} = 8.65$; Tukey's significant difference = $\sqrt{26/5}(4.041) = 9.21$ ($P < 0.05$). Treatment 1 is significantly lower than the others. (b) $\hat{\sigma}^2_{RCBD} = 26$, $\nu_{RCBD} = 10$, $\hat{\sigma}^2_{CRD} = [2239+10(26)]/14 = 178.5$, $\nu_{CRD} = 14$, $E_{RCBD:CRD} = (11/13)(17/15)(178.5/26)100\% = 658\%$.

8.3.   (a) Dose: $f = 13.18$ ($P < 0.01$); method: $f = 9.79$ ($P < 0.01$); interaction: $f = 0.40$ (not sign); means and standard errors: 57.6 ± 14.3(10 µg), 131.2 ± 14.3(20 µg), 157.6 ± 17.6 (iv), 67.7 ± 17.6 (im), 57.8 ± 17.6 (sc). (b) $ss_{NA} = 17,552$, $f = 5.61$ ($P < 0.05$); significant nonadditivity for animals and treatments. May need to group homogeneous animals in future experiments. (c) $f = 2.705 < f_{0.10,1,25} = 2.92$. Evidence is not strong for dependence of variance on level of response.

8.5.   (a) Parallel response (no interaction of dose and preparation): $f = 0.428$ (not significant); preparations: $f = 0.808$ (not significant; doses: $f = 20.21$ ($P < 0.01$). (b) $\hat{\sigma}^2_{CROSS} = 37.42$; $\hat{\sigma}^2_{CRD} = [1043.50 +12(37.42)]/15 = 99.50$; $(99.50/37.42)(4 \text{ rabbits}) = 11.$ rabbits.

8.7.   (a) Dose $ss = 179.06$, $ms_E = 0.454$, $f = 131.47$ ($P < 0.01$), 12 observations per dose; linear $ss = [-3(8)-1(29)+1(47)+3(71)]^2/\{(12)[(-3)^2 +(-1)^2+1^2+3^2]\} = 178.54$; $f = 393.25$ ($P < 0.01$); nonlinear $ss = 179.06. -178.54 = 0.52$ (2 df), $f = 0.573$ (not significant). (b) $\hat{\sigma}^2_{balanced} = 0.454$; $\hat{\sigma}^2_{random} = [1.56+33(0.454)]/36 = 0.4595$; $E = (0.4595/0.454)100\% = 101\%$. Random order (ignoring period effects) would be as good.

8.9.   (a) $\hat{\tau}_A - \hat{\tau}_B = 0.136\pm1.81$ ($ms_E = 19.66$), $t = 0.075$ (not significant). (b) Crossover (first 2 periods): $f = 1.411$ ($P < 0.25$), ($ms_E = 1.5968$), $\hat{\tau}_A - \hat{\tau}_B = 0.433\pm0.36$, $E_{CROSS:double\ reversal} = (19.66/1.5968)100\% = 1231\%$. [Note: If $E_{CROSS:CRD} = 1025\%$, then $E_{DR:CRD} = (1025/1231)100\% = 83\%$, i.e., completely randomized better than double reversal.]

8.11.   (a) Adjusted means: (A) 0.538 ± 0.022, (B) 0.727 ± 0.022, (C) 0.841 ± 0.022, (D) 0.896 ± 0.022, (E) 0.827 ± 0.022; $f = 39.76$ ($P < 0.01$). (b) $\hat{Y} = 0.766 +0.0747\xi'_{i1}-0.411\xi'_{i2} = -1.476+0.0415x-0.000183x^2$; $x_s = -0.415/ [2(-0.000183)] = 113\%$ of the methionine content of milk. (c) $(20/22) \cdot (42/40)(0.0595/0.0025)(20 \text{ calves}) = 454$ calves.

8.13.   (a) $f = 0.21 < f_{0.05,3,21} = 3.07$; approximate test of homogeneous variance: $f_{max} = 2.210 < f_{max,0.25,4,7} = 4.41$; approximate test of extreme correlations: $z = 1.50 > z_{0.9} = 1.28$. Some evidence against homogeneous correlation. (b) Tukey's test (assuming homogeneous variance-covariance): minimum significant difference (80% confidence) = $q_{0.20,4,21}\sqrt{ms_E/r}$ = $2.909\sqrt{0.24/8}$ = 0.503; maximum sample difference = 1.994-1.758 = 0.146 (no significant difference). Scheffé's test (modified for heterogeneous variance-covariance) would provide even less convincing evidence.

8.15. (a) Diets (A), f = 1.410 (no significance); ages (B), f = 15.963 $(P < 0.01)$; AB, f = 4.058 $(P < 0.01)$. B and AB are significant, even with conservative critical values. Homogeneity of variance from dose to dose at a given age: f_{max} = 1.97, 3.29, 12.75 $(P < 0.05)$, 2.90, 3.52. Diet 3, age 3, has low variance (by chance?); otherwise relatively homogeneous. Homogeneity of variance from age to age for a given diet: f_{max} = 8.87 $(P < 0.25)$, 8.84 $(P < 0.25)$, 19.08 $(P < 0.05)$. Some evidence for heterogeneity. Use conservative critical values. (b) Average variance (ms_E) for each age: 0.01070, 0.02233, 0.03343, 0.02307, 0.07827. Tukey's minimum significant difference (95%) between 2 diets at a given age: $q_{0.05,3,15}\sqrt{ms_E/r}$ = 0.154, 0.224, 0.276, 0.228, 0.419. Age 4: diet 2 > diet 1; age 5: diet 3 > 2. (c) Cycles 1-4 vs. age 5, within each diet: $\bar{q}$ = 1.161, 0.468, 2.659, with 95% confidence interval $\pm\sqrt{(b-1)f_{0.05,1,15}\hat{V}[\bar{q}]}$ = $\pm\sqrt{4(4.54)\hat{V}[\bar{q}]}$, where $\hat{V}[\bar{q}]$ = $[\sum_{i=1}^{4} s_i^2 + (4)^2 s_5^2$ $+2\sum_{i<i'}^{3}\sum s_{ii'}^{4} - 4(2\sum_{i=1}^{4} s_{i5})]/r$ = 0.2333, 0.0933, 0.2920. Intervals: 1.16 ±2.06, 0.47±1.30, 2.66±2.30 $(P < 0.05)$. Longer cycles at the oldest age for high alcohol diet.

8.17. (a) Surgery (A): f = 19.74 $(P < 0.01)$; X ray (B): f = 0.08 (not significant); AB: f = 0.03 (not significant); age (C): f = 49.61 $(P < 0.01)$; AC: f = 2.34 (not significant); BC: f = 4.65 $(P < 0.01)$; ABC: f = 8.02 $(P < 0.01)$. C, BC, and ABC are significant, even with conservative critical values. Homogeneity of variance from treatment to treatment at a given age: f_{max} = 11.07 $(P < 0.05)$, 5.17, 1.99, 5.31, 3.63. Homogeneity of variance from age to age for a given treatment: f_{max} = 10.20 $(P < 0.05)$, 6.34, 3.29, 4.72. High sample variance for A_1B_1 at 1/2 month is responsible for the two significant results. If chance alone is operating, ordinary procedures for specific comparisons may be reasonably valid. (b) Ages 1/2, 1, 2 vs. 3, 5 for each treatment combination: $\bar{q}$ = 177.8, 240.1, 66.7, 204.2, with 99% confidence interval $\pm t_{0.005,128}\sqrt{\sum_i c_i^2(ms_E/r)}$ = 68.0. Lymphocytes are significantly higher at 3 and 5 months of age than at earlier ages, but the difference is least pronounced when thymectomy is performed without irradiation (A_2B_1).

8.19. (a) $Y_{ijkl} = \mu + \alpha_i + D_{(i)j} + \beta_k + (\alpha\beta)_{ik} + (D\beta)_{(i)jk} + \gamma_l + (\alpha\gamma)_{il} + (D\gamma)_{(i)jl} + (\beta\gamma)_{kl} + (\alpha\beta\gamma)_{ikl} + (D\beta\gamma)_{(i)jkl} + E_{(ijkl)}$, where α, D, β, and γ are effects of treatment, rat, age, and minutes after feeding. Treatment (A): f =

12.38 (P < 0.05); age (B): f = 3.08 (not significant); AB: f = 8.58 (P < 0.05); minutes (C): f = 84.75 (P < 0.01); AC: f = 36.61 (P < 0.01); BC: f = 13.14 (P < 0.01); ABC: f = 3.30 (not significant). Average effect of age was not significant, but it interacts with treatment and time after feeding. (Inferences unchanged by using conservative critical values.) Variance is relatively homogeneous from treatment to treatment, age to age, and time to time, except for large sample variance in the $A_2B_2C_2$ combination that is largely caused by rat 5 (possible outlier). (b) Dunnett test (0 vs. 20 or 40 minutes, within each AB combination): 95% minimum significant difference =

$t_{D,0.05,2,12}\sqrt{2(ms_{E_{bc}}/r)}$ = 11.59. Significant rise in glucose at 20

minutes in control rats was not seen in treated rats. Glucose was lower at 40 minutes at the older age for control rats; at the younger age (or both ages?) for treated rats.

8.21. (a) Y_{ijk} = $\mu+\alpha_i+D_j+(\alpha D)_{ij}+\beta_k+(\alpha\beta)_{ik}+(D\beta)_{jk}+(\alpha D\beta)_{ijk}+E_{(ijk)}$, where Y = body weight at a given age, μ = mean body weight, α_i = effect of treatment, D_j = effect of block (random), β_k = effect of age, $(\alpha\beta)_{ik}$ = interaction of treatment with age, $(D\beta)_{jk}$ = interaction of block with age (error$_b$), $(\alpha D\beta)_{ijk}$ = interaction of block with treatment and age (inseparable from $E_{(ijk)}$, residual error) or error$_{ab}$. (b) Treatments (A), 5 df; blocks (D), 11 df; AD = error$_a$, 55 df; age (B), 3 df; AB, 15 df; DB = error$_b$, 33 df; $ADB+E$ = error$_{ab}$, 165 df. $H:\alpha_i$ = 0, f = ms_A/ms_{AD}; $H:\beta_k$ = 0, f = ms_B/ms_{DB}; $H:(\alpha\beta)_{ik}$ = 0, f = ms_{AB}/ms_{ABD}.

8.23. (a) Y_{ijklm} = $\mu+\alpha_i+\theta_{(i)j}+D_{(ij)k}+\rho_l+\beta_m+(\alpha\beta)_{im}+E_{(ijklm)}$, where Y = sperm count for 1 boar in 1 period, μ = mean sperm count, α_i = effect of zinc treatment, $\theta_{(i)j}$ = effect of square (pair) within treatment, $D_{(ij)k}$ = random effect of boar within square, ρ_l = effect of period, β_m = effect of ejaculatory frequency, $(\alpha\beta)_{im}$ = interaction of zinc and ejaculatory frequency, and $E_{(ijklm)}$ = residual error. (b) Treatments (zinc = A), 1 df; squares/A, 2 df; boars/square (D), 4 df; periods, 1 df; ejaculatory frequency (B), 1 df; AB, 1 df; error, 5 df. $H:\alpha_i$ = 0, f = ms_A/ms_D; $H:\beta_m$ = 0, f = ms_B/ms_E; $H:(\alpha\beta)_{im}$ = 0, f = ms_{AB}/ms_E.

8.25. (a) Dry matter (A): f = 0.042 (not significant); feeding level (B): f = 6.839 (P < 0.05); AB: f = 3.100 (not significant). Calves fed ad libitum gained faster. Some (but not strong) evidence was produced to show that the ad libitum advantage was stronger for 10% dry matter than for 15% (AB interaction). (b) 0.061±0.038; 0.052±0.037 (essentially the same precision for B and A means, in this case).

8.27. (a) q = 8.756 > $\chi^2_{0.10,4}$ = 7.779 (90% confidence that variance-covariance matrix is not homogeneous). (b) T^2 = 7.582 < $T^2_{0.05,2,5}$ = 17.361

(period means do not differ significantly); 95% intervals: (1 vs. 3), 13.2±44.9; (1 vs. 5), −8.8±76.9).

8.29. (a) q = 16.35 < $\chi^2_{0.05,10}$ = 18.307. (b) Treatments: U = 0.7642 > $U_{0.05,4,1,16}$ = 0.5055 (not significant); periods: U = 0.2025 < $U_{0.05,4,3,16}$ = 0.2429 (significant); interaction: U = 0.5502 > $U_{0.05,4,3,16}$ = 0.2429 (not significant).

CHAPTER 9

9.1. (a) $\hat{\beta}_0$ = 81.22, $\hat{\beta}_1$ = 1.97, $\hat{\beta}_2$ = 0.22, $\hat{\beta}_{11}$ = −3.93, $\hat{\beta}_{22}$ = −1.38, $\hat{\beta}_{12}$ = −2.22. (b) Stationary point at x_{1s} = 0.30, x_{2s} = −0.16; characteristic roots λ_1 = −4.35, λ_2 = −0.96, indicating a maximum more sensitive to changes in methionine than histidine; w_1 = 0.936$(x_1-0.30)$+0.352· $(x_2+0.16)$, w_2 = 0.352$(x_1-0.30)$−0.936$(x_2+0.16)$.

9.3. (a) $\hat{Y}$ = 66.3889−1.4400x_1−2.2812x_2−1.0950x_3−11.3561x_1^2−13.6798x_2^2−3.4972x_3^2 +9.1000x_1x_2+0.6075x_1x_3+0.8125x_2x_3. (b) Stationary point at x_{1s} = −0.1198, x_{2s} = −0.1286, x_{3s} = −0.1819; characteristic roots λ_1 = −3.4414, λ_2 = −7.8765, λ_3 = −17.2153, indicating a maximum most sensitive to changes in equilibration time (in the units coded);

$$
\begin{bmatrix} w_1 \\ w_2 \\ w_3 \end{bmatrix}
\begin{bmatrix} 0.81678 & 0.07572 & 0.99378 \\ 0.78586 & 0.60839 & -0.11084 \\ 0.61298 & -0.79003 & 0.01006 \end{bmatrix}
\begin{bmatrix} x_1+0.1198 \\ x_2+0.1286 \\ x_3+0.1819 \end{bmatrix}
$$

9.5. Full replicate of 2^4 = 16, plus 8 axial points (at ±2), plus r_2 = 12 center points; or half replicate of 2^5 = 16, plus 10 axial points (at ±2), plus r_2 = 10 center points.